T4-AIL-439

Water Resources Development and Management
Series Editors: Asit K. Biswas and Cecilia Tortajada

Asit K. Biswas
Cecilia Tortajada
(Eds.)

Impacts of Megaconferences on the Water Sector

With 25 Figures and 47 Tables

Editors
Prof. Asit K. Biswas
Third World Centre for
Water Management
Avenida Manantial Oriente 27
52958 Los Clubes, Atizapan
Mexico
akbiswas@thirdworldcentre.org

Cecilia Tortajada
Third World Centre for
Water Management
Avenida Manantial Oriente 27
52958 Los Clubes, Atizapan
Mexico
ctortajada@thirdworldcentre.org

ISBN: 978-3-540-37223-3 e-ISBN: 978-3-540-37224-0

DOI 10.1007/978-3-540-37224-0

Water Resources Development and Management ISSN: 1614-810X

Library of Congress Control Number: 2008942377

Cover design: Integra Software Services Pvt. Ltd.

Printed on acid-free paper

9 8 7 6 5 4 3 2 1

springer.com

Preface

Water-related issues have received increasing global attention during the past decade, certainly more than what existed a decade earlier when global interest in water issues was lukewarm at best. A good indication of this lack of global interest can be gleaned from the proceedings of the United Nations Conference on Environment and Development, held in Rio de Janeiro, Brazil, in June 1992, during which not even a single President or Prime Minister raised in any serious fashion the issue of water in their plenary statements. The Prime Minister of Bangladesh did mention water, but it was exclusively in terms of the country's problems with India because of the lack of agreement on the allocation of the flow of the Ganges River.

By the end of 1990s, and during the early part of the 21st century, the situation did change significantly in many ways. The global interest on water issues increased, and there were numerous discussions at national and international fora on issues like how water would become the most critical global resource issue of the 21st century just like energy was during the early 1970s; how the world would run short of water as a result of which developments in many parts of the world would be seriously constrained, and how there would be increasing conflicts between nations over the use of transboundary water bodies which would lead to water wars. In spite of these discussion and the fact that the media became very interested in the idea of global water scarcities or water wars, the real situation continues to be somewhat mundane. All the current serious analyses indicate that even in the driest parts of the world, like the Middle East and the North African region, their present and future water problems can be solved by using available knowledge, economic instruments, technology and management techniques. Just like the neglect of water issues in the international socio-political agenda of the late 1980s and early 1990s is difficult to understand or justify, the scare-mongering of the recent years in terms of a looming global water crises of unprecedented proportions or water wars are equally simplistic and unjustifiable.

There is no question that if the world faces a serious water crisis in the coming decades, it will not be due to lack of physical scarcities of water, but because of poor management of this resource. In nearly all developed and developing countries, water management practices and processes continue to be inefficient and suboptimal. They can mostly be improved very substantially. Herein lies the crux of the problem, which the development professionals in general, and the water professionals in particular, have failed to appreciate, as have most international institutions.

The water problems the world is facing are multidimensional in nature, their complexities are increasing with time, they often vary with time and space, and there are no global solutions for a highly heterogeneous and rapidly changing world that will be equally applicable to all the countries. Thus, the idea that people from different parts of the world, from different sectors and disciplines, and with different interests, could meet periodically to discuss the extent and nature of the problems, their potential solutions, successes, failures and constraints to implement the proposed solutions and anticipate the future water problems and solutions of the world because of changing population dynamics, migration, societal expectations and aspirations, management practices, and increasing globalization is a very attractive concept. Conceptually at least, people could meet periodically which would facilitate North–North, North–South and South–South knowledge, experience and technology transfer. Prima facie, this appears to be a very effective option. At such gatherings, one could find out which solutions have actually worked, where, why and under what enabling conditions, and then consider their potential replicabilities to solve the water problems one is facing elsewhere. Furthermore, at least conceptually, it should be possible to find out which solutions are not working, where and why, irrespective of their earlier promises. These types of discussions and knowledge dissemination should have considerable potential to improve water management in different parts of the world.

Accordingly, megaconferences in the water sector, where thousands of participants from different parts of the world can meet to exchange ideas, views and experiences is, at least on the surface, a very attractive concept. During the last two decades, the number of megaconferences in the water-related sectors has proliferated. The question therefore is, are these megaconferences improving water management practices and processes so that objectives like economic efficiency, improved quality of life, poverty alleviation and environmental conservation are being better fulfilled compared to what may have been the case if they had not taken place? Or have these, as some critics have claimed, become "Woodstock" of water, where a good time is had by all under the pretext of a conference, where one's expenses are covered by someone else?

Unfortunately, none of the water-related megaconference has ever been evaluated in terms of its usefulness, cost-effectiveness and overall impacts to give any definitive answer to the above questions. As a result, how useful they have been to promote efficient water management in different parts of the world is simply unknown. Equally, while reasonable estimates can be made of the costs of convening these megaconferences, including their opportunity costs, their benefits and overall impacts are mostly unknown and never estimated. Whatever information may be available on the extent of their benefits, or nature of their beneficiaries, is primarily anecdotal in nature, and thus of very limited use.

As our analyses show, all is not well with the global megaconferences in terms of their outputs, impacts or cost-effectiveness. There is near unanimity among the water professionals surveyed from all over the world that most of the megaconferences are having only marginal impacts on the water sector. There is no question that an overwhelming majority of our respondents felt that the organizational processes and the structures of these meetings need to be vastly improved to substantially increase their outputs and impacts. However, there was no such unanimity when the issue came to how the organizational processes and structures should be modified to ensure significantly higher levels of impacts, and/or what are the alternatives to the megaconferences.

In order to fill this gap, the Third World Centre for Water Management, with the financial support of the Sasakawa Peace Foundation of the United States, carried out a study on the impacts of the megaconferences on the water sector. The present book is the result of this evaluation. On behalf of our Centre, we would like to express our appreciation to Mr. Keiji Iwatake, Director of Sasakawa Peace

Foundation of the United States, and Dr. Seki Akinori and Dr. Sim-Yee Lau, President and Programme Advisor of the Sasakawa Peace Foundation, respectively.

The first-ever attempt to seriously evaluate the impacts of megaconferences on the water sector on a global basis is a complex task under the best of circumstances, not only for the methodological issues involved but also for the reliable information obtained on which such assessments could be based. We are thus most grateful to eminent international water experts like Gourisankar Ghosh, John Lane, Anthony Milburn, Morris Miller and Robert Varady who accepted our invitations to prepare think pieces in terms of their own personal assessments of the impacts of these megaconferences on the water sector. Their assessments are included in the book. Together, all these personal assessments cover Africa, Asia, Europe, Latin America and North America.

The book also includes specific country or regional assessments. These were undertaken by equally eminent water experts like Mr. ATM Shamsul Huda (Bangladesh), Prof. Mikiyasu Nakayama (Japan), Dr. C.D. Thatte (India), Dr. Anthony Turton (South Africa) and Prof. Olli Varis (Scandinavia). Their assessments are much appreciated.

The thinkpieces and country/regional analyses were discussed at a special invitation-only workshop at the Asian Institute of Technology, Bangkok. All the participants were invited in their personal capacities for a free and frank discussion of the commissioned papers and analyses. Following the discussions at Bangkok, the authors finalized their papers which are included in the present book. On behalf of the Third World Centre for Water Management, we would like to thank all the authors, all of whom participated at the Bangkok workshop, and also the other specially invited guests for their constructive comments and contributions to the meeting.

Last but not least, we would like to express our appreciation to Andrea Lucia Biswas Tortajada for helping us with the analysis of the global questionnaires, and Thania Gomez for formatting the book and other assistance in terms of the preparation of the final manuscript. Andrea Lucia presented her analysis of the questionnaire survey at the Stockholm Water Symposium for which she received a special award.

We are confident that the publication of this book will significantly contribute to increased discussions of this issue which has been totally neglected by the water profession thus far.

Asit K. Biswas
President, Third World Centre for Water Management, Mexico;
Senior Policy Advisor, Ministry of Environment, Aragon, Spain; and
Distinguished Visiting Professor, Lee Kuan Yew School for Public Policy, Singapore

Cecilia Tortajada
Vice President, Third World Centre for Water Management, Mexico;
Scientific Director, CIAMA, Zaragoza, Spain; and
Visiting Professor, Lee Kuan Yew School for Pubic Policy, Singapore

Contents

List of Contributors

Biswas, Asit K., President, Third World Centre for Water Management, Avenida Manantial Oriente 27, Los Clubes, Atizapán, Estado de México, 52958, México; Senior Advisor, Ministry of Environment, Aragon, Spain, and Distinguished Visiting Professor, Lee Kuan Yew School for Public Policy, Singapore

Drackner, Mikael, Representative, Svalorna LA, Bolivia, Victor Sanjinés 2866, La Paz, Casilla 6092

Earle, Anton, Project Director, Stockholm International Water Institute (SIWI), Drottninggatan 33, SE-11151, Stockholm, Sweden

Furuyashiki, Kumi, Department of International Studies, Graduate School of Frontier Sciences, University of Tokyo, 5-1-5 Kashiwanoha, Kashiwa, Chiba 277-8583, Japan

Ghosh, Gourisankar, Chair, Technical Expert Group, Rajiv Gandhi National Drinking Water and Sanitation Mission, Government of India, New Delhi, India

Huda, Shamsul ATM, Former Secretary, Ministry of Water Resources, House No. 8, Road No. 90, Gulshan Circle 2, Dhaka, 1212, Bangladesh

Iles-Shih, Matthew, MD-MPH Student, School of Medicine, Oregon Health and Science University, 3181 SW Sam Jackson Park Road, Portland, OR 97239-3098, USA

Lane, Jon, Executive Director, Water Supply and Sanitation Collaborative Council, International Environment House, 9 Chemin de Anemones, CH 1219 Chatelaine, Geneva, Switzerland

Milburn, Anthony, Former Executive Director, International Water Association, Ambourne Environments, 34 Church Meadow, Surbiton, Surrey KT6 SEW, UK

Miller, Morris, Deputy Secretary-General, UN Conference on New and Renewable Sources of Energy, 81-263 Botanica Private, KIY 4P9, Ottawa, Canada *(Deceased)*

Nakayama, Mikiyasu, Department of International Studies, Graduate School of Frontier Sciences, University of Tokyo, 5-1-5 Kashiwanoha, Kashiwa, Chiba 277-8583, Japan

Renko, Terhi, Water Specialist, Pöyry Environment Ltd., P.O. Box 50, FIN-01621 Vantaa, Finland

Thatte, C.D., Former Secretary Ministry of Water Resources, India, C-16 Parnali Society, Damle Path, Law College Road, Pune 41104, India

Tortajada, Cecilia, Scientific Director, International Centre for Water, Zaragoza, Spain, Vice-President, Third World Centre for Water Management, Avenida Manantial Oriente 27, Los Clubes, Atizapán, Estado de México, 52958, México; Visiting Professor, Lee Kuan Yew School for Public Policy, Singapore

Turton, Anthony, Unit Fellow, Water Resource Competence Area, Council for Scientific and Industrial Research (CSIR), P.O. Box 395, Pretoria, 0001, Republic of South Africa

Varady G., Robert, Deputy Director and Research Professor, Udall Center for Studies in Public Policy, University of Arizona, 803 East First Street, Tucson, AZ 85719, USA

Varis, Olli, Research Professor, Helsinki University of Technology, Water Resources Laboratory, Address: P.O. Box 5200, 02015 Espoo, Finland

PART I: OVERALL ANALYSIS

1 Impacts of Megaconferences on Global Water Development and Management

Asit K. Biswas

1.1 Introduction

The concept of megaconferences is not new. However, like most concepts, the approach has evolved considerably over time. The process was started during the early 1970s, somewhat inadvertently, by the United Nations, with its Conference on the Human Environment, in Stockholm, in 1972. This intergovernmental conference was initially proposed by Sweden to discuss existing and emerging environmental issues, including acid rain that was having significant adverse impacts on the ecosystems of the Swedish lakes and forests but which the Swedish authorities were unable to control because the atmospheric emissions originated from Germany and the United Kingdom. It was convened at a high decision-making level, and was instrumental in giving the nascent national and international environmental movements a major push. The success of the Stockholm Conference immediately spawned a new trend of megaconferences on priority global issues. The United Nations, shortly thereafter, followed through with a series of megaconferences on Population (Bucharest, 1974), Food (Rome, 1974), Women (Mexico City, 1975), Human Settlements (Vancouver, 1976), Water (Mar del Plata, 1977), Desertification (Nairobi, 1977), New and Renewable Sources of Energy (Nairobi, 1979), and Science and Technology for Development (Vienna, 1981).

Some of these megaconferences attracted more global attention than the others, and, not surprisingly, their global impacts varied widely. Generally, the earlier meetings, especially those held up to 1977, generated more international interest and impacts than the later ones. For example, by the time the Conference on New and Renewable Source of Energy was convened in Nairobi, in 1979, the global interest in megaconferences had waned very significantly, even though this meeting was on an important issue. Neither the Nairobi Energy Conference nor the Vienna Conference on Science and Technology left any visible footprint on the world.

The participation to these megaconferences was restricted only to governmental delegations and representatives of intergovernmental organizations. Sometimes, concurrent to a megaconference, there were parallel meetings in which the general public could participate. However, the number of participants at most of these parallel meetings was small. In addition, these parallel meetings had no impact on the discussions, conclusions and recommendations of the main events.

Exactly 20 years after these megaconferences, the United Nations decided to revisit those issues on which there were still considerable political interests. These included Environment (Rio de Janeiro, 1992), Food Security (Rome, 1994), Population (Cairo, 1994), Women (Beijing, 1995) and Human Settlements (Istanbul, 1996). In addition, during the 1990s, under the aegis of the United Nations, a framework convention on Desertification was also agreed to.

In retrospect, for a variety of reasons, water basically disappeared from the international political agenda during the 1980s and 1990s. For example, during these two decades, there was not even any serious discussion at any United Nations fora on the necessity or desirability of organizing a global consultation on water-related issues, 20 years after Mar del Plata, as was done for many other issues that were still considered to be important, i.e. food, population, women and human habitat.

Unfortunately, however, there was not only no serious and comprehensive review of the global water situation 20 years after the Mar del Plata megaconference, but also there was no serious discussion at any United Nations agency on the need, importance and relevance of organizing such a consultation. In retrospect, somehow, water simply no longer was considered to be a priority political issue by the international community during the 1990s.

A few people that were directly associated with the UN System (for example, Rodda, 1993) have argued that water was indeed an important issue during the UN Conference on Environment and Development, held in Rio de Janeiro in 1992, since the Chapter 18 of Agenda 21 was on water. Regrettably, however, several factors negate such delusions. First, very few water professionals from developing countries seriously participated at the Rio Conference, or during its long preparatory process, which were almost exclusively dominated by the officials from the national Environment ministries. Equally, the heads of states that were present during the Rio deliberations did not refer to water as an important environmental issue. The only exception was the Prime Minister of Bangladesh, who only referred to the complexities of the water sharing arrangements of the River Ganga with India. Not even one single head of state referred to the water problems the world was facing, or likely to face in the future, or the importance of good water governance. In addition, the Chapter 18, even though it was the longest chapter of the Agenda 21, had one of the poorest frameworks. It also lacked any serious intellectual or technical gravitas. In all probability, water developments all over the world would not have been any different at present, even if the Rio Conference had not taken place. Thus, the claim that Rio was a major global milestone for the water sector has absolutely no factual basis.

However, while the water part of Rio (Chapter 18) did not have any visible and long-term impact on the water sector, the environmental components of Rio most certainly had perceptible impacts on the water sector. In fact, shortly after Rio, countries like Brazil and Mexico put the water-related issues under the jurisdiction of the Ministry of the Environment, and the environmental aspects of water development started to receive much more attention in the water sector compared to the situation that existed before the Rio Conference took place.

These and other similar developments indicate that any objective and realistic assessment would have to conclude that water as a whole basically disappeared from the international political agenda during most of the 1980s and 1990s. Any objective assessment will indicate that water was not considered to be a priority global issue during this period.

More than three decades after the Mar del Plata Conference, it is important to objectively and constructively review the progress that has been made during this period in the water sector globally, especially in terms of successes, shortcomings and constraints. It is also necessary to assess realistically the water-related issues that the world is likely to face in the future (Biswas et al., 2009).

The main analysis in this chapter, on the evolution of megaconferences related to the water sector, starts from the Mar del Plata Conference of 1977 and continues until the end of the Third World Water Forum in Japan in 2003.

1.2 Mar del Plata in Retrospect

It is worthwhile to recall the main objective of the Mar del Plata Conference, which has so far been the only major and substantive water meeting that has ever been held at a high political intergovernmental level in the entire human history. Its objective was "to promote a level of preparedness, nationally and internationally, which would help the world to avoid a water crisis of global dimensions by the end of the present century." The Conference was to deal with "the problem of ensuring that the world had an adequate supply of good quality water to meet the socio-economic needs of an expanding population" (Biswas, 1978).

The expectations of the Mar del Plata, in the words of its remarkable Secretary General, Yahia Abdel Mageed, were as follows:

> It is hoped that the Water Conference would mark the beginning of a new era in the history of water development in the world and that it would engender a new spirit of dedication to the betterment of all peoples; a new sense of awareness of the urgency and importance of water problems; a new climate for better appreciation of these problems; higher levels of flow of funds through the channels of international assistance to the course of development; and, in general, a firmer commitment on the parts of all concerned to establish a real breakthrough so that our planet will be a better place to live in (Mageed, 1978).

Concurrent to the official UN Conference, there was also another meeting on water at Mar del Plata, which was primarily attended by academics, government officials and NGOs. Overall, the level of participation to this parallel meeting left much to be desired (less than 400 people attended), and the participants were mostly local. In addition, the discussions at this parallel meeting had no relation or impact whatsoever on the official deliberations that took place at the main Conference.

The Conference approved an action plan, which was officially called the Mar del Plata Action Plan. It was in two parts: recommendations that covered all the essential components of water management (assessment, use and efficiency; environment, health and pollution control; policy, planning and management; natural hazards; public information, education, training and research; and regional and international cooperation) and 12 resolutions on a wide range of specific subject areas.

A retrospective and objective analysis of the Conference achievements and its subsequent impacts on the world as a whole clearly indicates that it was more of a success than most of its ardent supporters believed at the time the meeting was held, or shortly thereafter. A comprehensive review of the Conference achievements, carried out in 1987, a decade after the event, indicated that it had numerous primary, secondary and tertiary impacts, which were for the most part significant and beneficial (Biswas, 1988). It was, undoubtedly, a major milestone in the history of water development during the second half of the 20th century.

The activities leading to the final Conference produced a wealth of new knowledge and information on various aspects of water management as well as country- and region-specific analyses. For the very first time, many developing countries produced detailed national reports on the availability and use of water, and also detailed assessments of planning needs and management practices (Biswas, 1978a). It is important to note that the massive documentation that was produced during its preparatory process and its results are still available. This is in sharp contrast to the subsequent water megaconferences like the Dublin Conference and the World Water Forums, whose documentation has been conspicuously absent.

Several developing countries, encouraged by the Mar del Plata event, put in motion processes to assess the availability and the distribution of their surface and groundwater resources, and existing and futures patterns of water demands and uses. Most developing countries not only have continued these activities, which were initiated during the preparatory process of the Water Conference, or shortly thereafter, but also have significantly strengthened them progressively during the past three decades. In retrospect, the activities leading to the Mar del Plata, the event itself, and its follow up activities, have contributed significantly more to water development than all the combined efforts made by the UN System as a whole either before the meeting or during the ensuing three decades. By any standard, it was a most remarkable achievement.

A major output of the Conference was the recommendation that the period 1980–1990 should be proclaimed as the International Water Supply and Sanitation Decade. The objectives of the Decade included that the world should be forcefully reminded that hundreds of millions of people did not have access to clean water and sanitation facilities, and accelerated political will and investments were essential to dramatically improve this unacceptable situation. Even the most confirmed critic of the international system will have to accept the fact that the Decade significantly changed the quality of life of millions of people all over the developing world in terms of access to clean water and sanitation. In spite of this remarkable progress, the task, of course, is far from complete for a variety of reasons including population growth and continued mismanagement of water resources. Equally, and most certainly, without the Water Conference, the progress in this area would have been much less than what can be observed at present.

In retrospect, the Water Conference also had an important impact on the United Nations System as a whole. During the 1970s, the rivalries between the various UN agencies working in the water area were intense. The work initiated by Secretary General Mageed on the potential modalities of collaboration between the various UN agencies went a long way to smoothen the interrelations between them. The intensive rivalries of the 1970s have gradually given way to extensive consultations but, unfortunately, still with limited real cooperation, between the agencies concerned during the past three decades. The absence of real cooperation has meant that the considerable synergy that water programmes of the various United Nations agencies could have produced has simply not been realized.

Viewed from any direction, the Mar del Plata has to be considered to be an important milestone in the entire history of water development and management. The main Conference itself, and the four regional meetings that preceded it, considered water management on a holistic and comprehensive basis, an approach that mostly became popular only a decade later.

Looking back, three areas should have received additional attention: financial arrangements, modalities for the implementation of the Action Plan and management of water resources shared by two or more countries. On the first issue of realistic financial arrangements needed to implement the Action Plan, regrettably this aspect has never received the attention it deserved at any UN megaconference thus far, starting from the Stockholm in 1972. Thus, not surprisingly, the ambitious Action Plans of these Conferences have never been satisfactorily implemented. Mar del Plata was no exception to this overall general situation. It was a systemic problem of the United Nations, which, most unfortunately, continues even to this day, some 36 years after the Stockholm Conference.

It is also a sad and regrettable fact that United Nations System has never critically analysed the efficiency of the processes used for organizing these megaconferences, their relative strengths and weaknesses, and the impacts of the final outcomes in terms of their implementation. Consequently, many of the mistakes made have continued from one conference to another. How the agreed to Action Plans could be realistically and cost-effectively implemented is one area that has consistently received inadequate attention during all the high-level UN megaconferences, and also in terms of serious discussions, both before, during and after these events, and within the UN System itself. The Water Conference was no exception to these practices.

For a variety of reasons, the management of international waters was not considered as comprehensively as it should have been at the Mar del Plata. In an objective and retrospective analysis of the Water Conference, its Secretary General candidly observed that this area was "not tackled satisfactorily at the Conference" (Mageed, 1982). He further suggested "a re-examination and re-evaluation of the Mar del Plata Action Plan" in order to "revive the spirit developed at the Conference and, hopefully, to give it a new vigour". Regrettably this excellent suggestion was never considered seriously by the UN System.

Even more unfortunately, the International Conference on Water and the Environment (ICWE), which was convened in Dublin, in January 1992, by the United Nations System as a prelude to the UN Conference on Environment and Development (UNCED), all but ignored the achievements and the impacts of the Mar del Plata, or the process that was used for its organization. So far as the organizers of the ICWE were concerned, not only they ignored totally the results, or implications and experience of the Mar del Plata Conference due to some inexplicable reasons, but also, most regrettably, it was a deliberate decision. This was because some of the people associated with the preparatory process of the Dublin meeting argued that they should bring "new blood" and "new ideas". In reality, not only any new usable ideas surfaced during the Dublin meeting but also the institutional memories of the United Nations System related to the Mar del Plata somehow disappeared completely during the preparatory process leading to the Dublin Conference and then at Dublin itself. This deliberate but most unfortunate decision to ignore the results and the contributions of Mar del Plata was one of the important reasons as to why the meagre impacts of the Dublin Conference were in sharp contrast to the significant achievements of the Mar del Plata. Some of these issues will be discussed next.

1.3 Absence of Water in the International Agenda after Mar del Plata

Fifteen years after the Mar del Plata, the world's leaders met at Rio de Janeiro, in June 1992 for the United Nations Conference on Environment and Development (UNCED). Most development and water professionals had expected that the UNCED would not only revive the spirit of the Mar del Plata but also would put water firmly in the international political agenda. Most unfortunately, however, exactly the reverse happened. Issues like climate change, biodiversity, deforestation and ozone depletion took the centre stage during the statements of the Presidents and the Prime Ministers at Rio: water was at best considered to be a "bit" player largely confined to the wings (Biswas, 1993).

The omission of water from the international political agenda, as was evident in Rio, and the subsequent developments are important but regrettable facts which the water profession needs to review very carefully. While some institutions and people have deliberately glossed over this unsatisfactory situation, the water profession can no longer ignore the facts and the reasons as to why the megameetings of Dublin and Rio failed so miserably to put water in the international political agenda, and also to contribute something substantial to accelerate the construction of much-needed water infrastructure and improve water management processes and practices in the developing world. Lessons should be learnt from such failures so that the same errors are not repeated in the future.

1.3.1 Failure of the Dublin Conference

The Dublin Conference was convened by the United Nations System and was expected to recommend appropriate sustainable water policies and action programmes for consideration by the UNCED. Unfortunately, it never achieved even these modest objectives. Its duration, only 4 months before the UNCED, was ill-conceived and ensured that it had at best marginal impacts on the deliberations at

Rio. Even if the Dublin Conference had come out with a single new idea or concept, which it did not, and had considered critical issues in terms of major programme initiatives, including how much would such programmes cost, where would the funds come from, and how and by whom would such new programmes be implemented, which again was basically ignored, there simply was not enough time available to effectively incorporate any ideas that could have come from Dublin into the Rio programme. In retrospect, the overall planning for the Dublin Conference left much to be desired, in terms of influencing the water-related agenda at Rio, ensuring promotion of efficient and equitable water management in the world, and harnessing new investment funds for the water sector.

Second, the Dublin Conference, for some incredible and inexplicable reasons, was organized as a meeting of experts and not as an intergovernmental meeting. This was in spite of the explicit advice given to the Secretariat by certain governments, notably Sweden, and several knowledgeable water experts, including the author, and the prevailing rules that governed the organization of the UN megaconferences. The distinction between a meeting of experts and an intergovernmental meeting is an important consideration in the context of any UN megaconference, since such conferences can *only* consider recommendations from intergovernmental meetings and *not* from an expert group meeting, as was the case for Dublin. Accordingly, and predictably, certain governments objected at Rio to any reference to the results of the Dublin Conference, irrespective of whatever may have been their importance or relevance, on procedural grounds, since it was *not* an intergovernmental meeting. It is still a mystery as to why the organizers of the Dublin Conference chose the route of expert group meeting approach, especially when they were very specifically warned that the results of such a process could not be considered at Rio because of the prevailing UN rules. Accordingly, and in all probability, the entire Chapter 18 of Agenda 21, which dealt with water, would have been very similar, irrespective of whether the Dublin Conference had ever been convened or not.

Thus, not surprisingly, during the Third Stockholm Water Symposium in August 1992, shortly after the Dublin Conference, the overall view of the participants was that the Dublin meeting was a failure, especially in terms of outputs and impacts, and that the water profession could not afford another similar major setback in the future (Biswas 1997).

During the 1990s it became "politically correct" for many national and international organizations to speak glowingly of the so-called Dublin Principles as if they were new and they would, by themselves, somehow contribute to efficient and equitable water development. Equally, it was often claimed that the four principles that were derived through an unplanned and *ad hoc* process were the most important ones in the field of water management. The four Dublin Principles were not even included in the Agenda 21 and, for the most part, are simply bland statements of the obvious, which even if they were implemented by a miracle, would *not* create the necessary enabling conditions for efficient water management. Most surprisingly, the principles did not refer to the fundamental objectives of water developments, like poverty alleviation, regional economic distribution or environmental conservation. The first two objectives, most surprisingly, were ignored at Dublin. This is in spite of the fact that no water development project is viable over the long-term if issues like equity and poverty are basically ignored. Not surprisingly, a scant few years after the Dublin meeting, its Principles became, at best, a brief footnote in the history of water management.

Furthermore, the Dublin Conference basically ignored the issue of water governance, which became a major consideration only a few years later.

An objective analysis may even indicate that in several instances Dublin may even have been a retrogressive step, especially if its results and impacts were compared to what were achieved at Mar del Plata. For example, one of the four Dublin Principles stated that water should be "recognized as an economic good". In contrast, 15 years earlier, Mar del Plata had specifically urged to "adopt appropriate pricing policies with a view to encouraging efficient water use, and finance operation cost with due regard to social objectives". This Principle was recommended not only for drinking and industrial uses but also for the irrigation sector. Dublin emphasized exclusively water as an economic good and, most surprisingly, ignored the historical fact that water has always been considered to be a social good. By ignoring totally social aspects of water, including equitable access, it created an unnecessary chasm between water as an economic good and a social good. More than a decade passed before the chasm could be bridged, and the world could return to the earlier paradigm that water is both a social and an economic good.

In addition, the so-called Dublin Principles are generalities, and at best could be considered to be good rhetoric and which mostly reflected the short-term political views of some people of that time. The four Principles are of limited value to developing countries which are searching for alternatives as to how best to formulate and implement efficient water management policies and programmes. Furthermore, no thought was given at Dublin as to how these vague principles could be operationalized by the decision-makers and water professionals in developing countries, or elsewhere for that matter. Now, some 16 years after Dublin, the die-hard supporters of the Dublin Principles have mostly disappeared, and the very few that are left have consistently failed to show how these Principles can be operationalized in the context of efficient water management in a complex but real world. Also, neither Dublin nor Rio has had any perceptible impact on the water sector which would not have occurred even if these two events had not taken place.

1.4 World Commission for Water

Past experiences indicate that world commissions are generally not easy to organize and manage. Even more difficult is to establish a World Commission that can produce something useful and worthwhile that could have lasting impacts. To its credit, the World Commission for Water did manage to assemble a very distinguished group of individuals, who willingly agreed to serve on the Commission in their personal capacities at a very short notice.

Right from the very beginning, the Commission had a very tight time schedule to organize itself and to produce a report within a period of a little over a year. The time element was critical since the Commission decided to undertake the exercise in a participatory manner that would include as many stakeholders as possible from different parts of the world. It also made a very special effort to engage women in all its discussions. The consultative process eventually encompassed thousands of individuals from all over the world, representing hundreds of institutions that were local, national, regional or global in nature, and both governmental and non-governmental. In terms of process, it was thus a unique and complex exercise. Never before in the entire history of water, such an exercise was ever attempted, let alone carried out.

The Commission reviewed the results of all the consultations and the discussions to produce a final report entitled: "A Water Secure World: Vision for Water, Life and Environment" (World Commission for Water in the 21st Century, 2000). The report was concise (only 68 pages), and written in a form that was easily understandable by anyone interested in water. Equally, since the Commission was independent, it managed to make several recommendations which may not have been possible through intergovernmental fora like those of the United Nations or the World Water Council where consensus rules the day.

The main thrusts of the report of the Commission can be summarized as promoting:

- holistic, systemic approaches based on integrated water resource management;
- participatory institutional mechanisms;
- full-cost pricing of water services, with targeted subsidies for the poor;
- institutional, technological, and financial innovations; and
- governments as enablers, providing effective and transparent regulatory frameworks for private actions.

The Commission believed that the above requirements will not be achieved until and unless attitudinal shifts occur, resulting in:

- mobilization of political will; and
- behavioural change by all.

The Commission recognized that much more work needs to be carried out so as to mobilize the necessary political will to implement its finding and recommendations.

According to the Commissioners, "the single most immediate and important measure" that they could "recommend is the systematic adoption of full cost pricing for water services." The report suggested that "an essential element will be to use targeted, time-bound subsidies to attract first class service providers who can be paid for the costs of their services and provide users with high quality services." The reasons for this recommendation were the following:

- free water leads to wastage and inefficient use;
- considerable resources are invested in the water and sanitation sectors in developing countries which were estimated at $30 billion per year;

- governments in developing countries could not even meet the existing investment demands for water services, let alone the very substantial requirements for the future; and
- limited public resources are devoted to public goods, specially environmental enhancement (for example, much of the wastewater produced in Africa, Asia or Latin America are now inadequately treated).

No reasonable person will disagree with the Commission's view that the day when water could be considered to be a free good that would be automatically provided by the governments at very low or zero costs is gradually, but most certainly, coming to an end. However, achieving water pricing will not be an easy task since there are simply too many vested interests in maintaining the current practices and also the *status quo*; too many dogmatic views which are often based on erroneous facts and/or understandings; and too many mind-sets that belong to the past. Equally, many people automatically assume that water pricing and making water management practices more efficient would mean automatic transfer of all the functions from the public to the private sector. This thinking was predominant during the Second World Water Forum in the Hague (hereafter referred to as the Hague Forum), and to a lesser extent at the Third World Water Forum in Japan (hereafter referred to as the Japan Forum).

This of course is not correct, since water utilities, irrespective of whether they are in public or private sector, will have to charge an appropriate price for water if universal access to clean drinking water and proper wastewater management is to be a reality. It should also be noted that both public and private sectors have their strengths and weaknesses. For example, one of the best examples of urban water management is the case of Singapore, where a public sector autonomous company has a superb record which compares very favourably with any water utility that is run by a private sector company anywhere in the world (Tortajada, 2006). Losses from the Singapore water system, currently about 4.6%, are now one of the lowest in the world. Equally, however, losses from many public sector managed water companies are also now running close to 40–60%, and in a few cases even up to 80% (Biswas, 2000). The performance variations of the public sector companies are simply far too diverse to draw any definitive conclusions. The performance of private sector companies has been equally variable. Accordingly, dogmatic views on the performances of the public or the private sectors are not universally valid. Each case should be considered on its own merits and constraints, and the prevailing local social, economic and institutional conditions. It should also be noted that the performances of the public or the private sector utilities may vary with time, sometimes quite significantly.

In the future, the main focus will unquestionably be to encourage public–private, public–public and public–private–civil society partnerships in many different forms, depending upon specific local conditions (Asian Development Bank, 2007). It should no longer be the continuation of the simplistic arguments like public sector versus the private sector, or whether water should be priced, or free, or heavily subsidized. Similarly, not a single public or private sector model, or water pricing model, will fit equally well to all countries, or even within one country. Furthermore, all these models will continue to evolve with changing social, economic and political conditions, public perceptions, technological developments and governance situations.

In retrospect, the overall impact of the report of the Commission (the author was a member) on the water sector has been minimal. Probably its main contribution was to encourage a large number of water professionals from different parts of the world to work together to develop visions for water. However, all these visions were poorly formulated in the sense that not even one of them identified how the world may look like in 2010, let alone a decade or more later (they were all supposed to be for 2020). Basically, most visions were similar, irrespective of the geographical areas concerned, and their social, economic, political and institutional conditions. In addition, they were very broad, general, linear and somewhat simplistic. They were also unusable in terms of formulating specific national policies that could be implemented.

Accordingly, and not surprisingly, all these visions were basically ignored by the governments of the countries concerned, and now have become primarily historical documents which are likely to be of very little use, either to the practitioners, or to the academics, or even to the people who formulated these visions. In addition, the regions selected often contained countries with very different levels of water availability, climatic and other physical conditions, management and technical capacities, institutional and legal frameworks, and varying levels of socio-economic development conditions. For example, the South Asian vision included both Bhutan and India. The visions and expectations of these two countries, one very large and the other very small, have to be very different. Not surprisingly, the South Asian vision was dominated by the large countries: unique features, accomplishments and expectations of a small country like Bhutan were mostly ignored. Such broad visions, developed exclusively by water professionals for very wide range of conditions, seldom have any practical value. This is a lesson, most unfortunately, the organizers of the megaconferences still have not learnt.

1.5 First and Second World Water Forums

The World Water Council organized the First World Water Forum in Marrakech, Morocco, 1997. The Council was new at that time, and was trying to carve out a role or niche for itself. The initial idea was to have a series of World Water Forums "with movers and shakers" of the water profession, somewhat similar to the World Economic Forums of Davos. This however proved to be an impossible dream for many reasons. First was the lack of finance. The Council had very limited funds which meant that a Davos type of forums was simply out of its reach. Second was the absence of good, long-term planning capacity, which simply did not exist. Third, the Council simply did not have the clout to bring together the movers and the shakers of the water community. Fourth, the different Council members had different views and agendas, sometimes polar opposite, which were simply not possible to be reconciled. In addition, political infighting within the Council by some members to push their own personal and institutional interests was intense which ensured that the decisions taken were seldom optimal, either for the Council or for the water profession as a whole. The decisions were reached with considerable tradeoffs between the various parties and interests which meant that they were often reduced to the lowest common denominator in order that these could be made acceptable to the Council as a whole.

The first Forum was attended by a few hundred water professionals. It was mostly a low-keyed affair, which consisted of continuous speech-making, from morning to late evening, with virtually no time for discussions and consultations. The Forum did come out with some recommendations, but how these were arrived at, or who prepared and promoted them, are still a mystery even to this day. These were mostly certainly not discussed in any fashion within the Forum framework.

On the positive side, the Forum did produce documentation containing some selected speeches (Ait Kadi et al., 1997). This simply did not happen for the Second, Third, or Fourth Forums.

The Second World Water Forum was organized in the Hague, the Netherlands, 17–22 March 2000, and its centrepiece was the Report of the World Commission on Water. The Forum was strongly supported organizationally and financially by the government of the Netherlands. According to the organizers, some 4,600 participants from all over the world participated in this event. It was thus a far bigger meeting, at least in terms of the number of participants, compared to the first Forum or the Mar del Plata Conference. However unlike the Mar del Plata, the Forum was sponsored by the World Water Council, a NGO, and not by an intergovernmental body like the United Nations. The large number of participants who attended the Forum confirmed a new global trend of the 1990s for the water sector. The important international roles played by the UN System during the pre-1980 period had started to decline, and this vacuum was then filled by new institutions like the World Water Council, Global Water Partnership, Stockholm Water Symposium and Singapore International Water Week. This general trend will probably continue well into the next decade. However, the UN System is now trying to carve out more visible roles for itself, especially within the context of the World Water Forums and the Stockholm Water Symposiums.

The Hague Forum was different from the Mar del Plata Conference at least in five important ways. First, unlike the continuous speech-making at the plenary sessions of the UN-sponsored megaconferences by the ministers and the other senior officials from all the countries present and by the heads of the intergovernmental organizations, the Hague Forum constituted over 100 sessions on a variety of topics, which included issues as diverse as water and energy, next generation of water leaders, water vision for Mexico, senior women water leaders, water and religion and business community (CEO) panel. Some of the sessions were well attended, but others had only 10–20 participants, including their organizers. Second, participation to the Forum was open to anyone who wished to participate and had the financial support to participate from their institutions or the donors. This was in contrast to Mar del Plata, where participation was very strictly restricted only to the official representatives of the governments and the appropriate intergovernmental organizations.

Third, framework and issues considered at Mar del Plata emerged from carefully structured and organized regional meetings, which had considerable technical and intellectual underpinnings. Also, its Secretariat commissioned think pieces from the leading international water experts. The Hague Forum was structured mostly on an *ad hoc* basis. Several hundreds of papers were presented at the Hague, without any real peer review or quality control. Thus, for the most part, the papers presented at the Hague or the Japan Forums were somewhat poor and were mostly of SOS (same old stuff) type (Biswas, 2006).

Fourth, the Mar del Plata Conference resulted in an Action Plan, which was accepted by all the governments that were members of the UN. The Hague and the Japan Forums did have Inter-Ministerial meetings which were restricted to senior government officials. Active participation of the ministers, however, was patchy at best, and very few of them took the process seriously. At both of these Forums, there were Ministerial Declarations that were not only very general but also they broke no new grounds, had any innovative idea or had any subsequent impact. A prominent Mexican journalist wrote that the results of the Inter-Ministerial meeting at the Hague was "like water: no odour, no colour and insipid". Finally, the two Forums gave no thought as to how the information that was presented could be disseminated to the interested water professional for possible use or implementation after the event was over. This was in sharp contrast to the Mar del Plata, where information dissemination, before, during and after the Conference was considered to be very important, and taken very seriously. A cynic, however, may argue that not much useful knowledge or results came out of the two Forums which were worth disseminating.

Much of the Hague Forum activities were conducted peacefully. There were heated discussions on a few issues, especially on privatization and large dams. For the most part, these were carried out in a civilized and democratic manner. There were some difficult moments, however. The Plenary Session was disrupted by a group of protesters who were protesting against the construction of a Spanish dam and privatization. Two of the protesters, a man and a woman, took off their clothes on the podium, and others chanted slogans or simply made loud noises, as a result of which the opening session had to be postponed. The "colourful" disruptions were obviously planned carefully well in advance, and ensured that their activities received extensive global media coverage, but because of wrong reasons. The proceedings finally restarted after a courageous personal intervention by the Prince of Orange.

Similarly, the session on water and energy, which was specially organized for the World Water Council to review the linkages between water and energy policies, was hijacked by a small group of 3–4 anti-dam activists from Narmada Bachao Andolan and International Rivers Network, who were interested only in a single issue (no large dams should be built anywhere in the world irrespective of their needs and benefits), which had nothing much to do with the main focus of the session. They unfurled banners, and their disruptions ensured that any civilized discussion on the focus of the session was impossible. This was indeed most regrettable since the water profession had basically ignored energy in the past, even though water and energy policies and uses are closely interlinked. These interlinkages all over the world have steadily increased since the year 2000, and are likely to increase even further in the future. Fortunately, the security in the Forum was increased very significantly after the first day, and this effectively eliminated unwarranted disruptions. Similar problems, very fortunately, did not happen during the Japan and the Mexico Forums.

Significant credit for the independence of the Forum must be given to the Dutch Government, who ensured that it remained a public event, where people could express their views and opinions without any governmental interventions or interference. Accordingly, when the National Water Commission of Mexico formally asked the Forum organizers to "modify the programme" so that *only* the "officially designated representative" of that country could present the official "vision" of the country, instead of the representatives of the Mexican civil society as was planned for the Forum, the request was politely but firmly declined by its Dutch organizers. This is a most welcome step that simply would not have been possible, had the Forum been organized under the aegis of the United Nations, or other similar intergovernmental institutions.

1.6 Bonn Conference and Johannesburg Summit

The main global water megaconference after the Hague Forum was the Freshwater Conference at Bonn, held in December 2001, which was expected to send a message on water to the World Summit on Sustainable Development that was later held in Johannesburg, in South Africa, in August/September 2002. Like its precursor, the Dublin Conference, which was expected to send a similar message to the Rio meeting, the results of the Bonn Conference look even weaker now compared to the Dublin discussions. Not only did it not break any new ground in terms of ideas, targets, investments or programmes, most of the discussions were equally of SOS type and some times even grossly out of date. In fact, a cynic may be excused for arguing that many of the Bonn statements have been heard repeatedly during the previous two decades! Except for the discussion on corruption, "political correctness" was the order of the day! Thus, and not surprisingly, the so-called "Bonn keys" simply disappeared from the collective memory of the water professionals within less than one year of the consultation!

The Ministerial Declaration of Bonn was equally vague and insipid as the Hague or the Kyoto declarations. In addition, the Bonn Declaration stands out for its stark neglect of the issue of the water requirements for the agricultural sector. This is in spite of the fact that agriculture is the main user of water, and water use for food production is a major consideration for the world as a whole. The primary focuses were on water supply, sanitation and water quality. This highly skewed outcome was most probably due to the interests of the organizers of the Bonn discussions. One can legitimately argue that the Bonn discussions focused more than 75% of their attention to less than 25% of the global water problems!

So far as the Johannesburg Summit is concerned, its overall impact on the water sector has been somewhat amorphous. On the positive side, certainly significantly more water professionals participated in this event, compared to the Rio Conference. However, this participation occurred only outside the framework of the intergovernmental discussions, where the main issues were discussed and the actual decisions were being taken. Thus, the overall impact of a larger non-governmental participation was mostly marginal on the final conclusions and declarations of this Summit. Realistically, the conclusions and recommendations on water-related issues would have been very similar irrespective of whether the non-governmental discussions had taken place or not. The Summit also broke no new grounds in the area of water, nor did it spawn any new definitive programme on water, or bring any new additional investment funds to the water sector. It reiterated the water-related Millennium Development Goals, and added one in the area of sanitation, which was a most positive development.

The global views on the achievements and impacts of the Johannesburg Summit have not been auspicious. Consider the following headlines from prestigious media from different parts of the world on the results of this Summit:

World Summit falls flat – *Asahi Shimbun*, Japan
Dialogue of the Deaf – *Daily Telegraph*, London
Big Agenda, little action – *International Herald Tribune*, Paris
A long way to go for little success – *Financial Times*, London
The bubble-and-squeak summit – *The Economist*, London
Was the sustainable summit a wash out? *The Economist*, London

These headlines probably reflect accurately its overall impacts on the water sector as well.

1.7 Third World Water Forum

The Third World Water Forum was held in Japan in March 2003. According to the official statistics, this Forum attracted significantly more than four times the number of participants compared to the Hague Forum. It had nearly three times the number of sessions. However, as any perceptive observer of the Forum may have noted that if there were so many participants, certainly significantly more than one-third were NOT present during the actual discussions on any day at Kyoto, Osaka and Otsu. Furthermore, whereas the Second Forum was held in one city, the Third Forum was held concurrently in three cities, Kyoto, Osaka and Otsu, which contributed to high levels of fragmentation. The large number of participants and sessions, spread over three cities, meant that no single participant or institution, including the Forum organizers, had a clear and overall view of what was happening during the Forum, and what, if any, were the main messages that came out of these discussions. Whereas the Second Forum had the binding thread of the Report of the World Commission on Water, the Third Forum basically constituted a mixture of some 350 independent sessions, which were impossible to integrate. It was simply an impossible task to distil the overall messages from all these diverse sessions in three locations.

Like the Second Forum and the Bonn Conference, the Third Forum also had an Inter-Ministerial meeting. The Ministerial Declaration was equally bland as the other two meetings, and has had no impact on water management and development practices in the world. The draft Ministerial Declaration was initially formulated primarily by the Japanese Ministry of Foreign Affairs, without adequate consultations with the national governments before the meeting took place. Some governments did take the Declaration seriously and sent detailed comments, which were mostly ignored. The whole process to prepare the Declaration was somewhat opaque and non-participatory. Equally, ministers present did not take the Declaration seriously in terms of its possible implementation later.

A special aspect of the Third Forum was a very genuine and praiseworthy effort by the Japanese organizers to ensure good and real participation by the water professionals as a whole in the event. There was a very genuine attempt by the Japanese Secretariat to make the Forum inclusive. This ensured that institutions and individuals could at best make marginal attempts to promote their interests and agendas. This, must regrettably, was not the case for the Fourth World Water Forum in Mexico.

Even though the Japan Forum turned out to be an expensive event, its impacts on the water sector have been somewhat marginal. In retrospect, it can probably be best described as a large "water fair", with large number of participants.

Based on these results, it is essential for the water profession to critically and objectively assess the impacts and the cost-effectiveness of the various major water-related global megaconferences. The existing implicit thinking that the number of people, or countries that participated in a megaconference, or the total money spent, can no longer be considered to be important, or even relevant, indicators of their success.

1.8 Comparison of Three Forums

So far as the Marrakech, the Hague and the Japan fora are concerned, an objective evaluation indicates that the Marrakech Forum had only speeches but no discussion. In contrast, the Hague and the Japan fora discussed numerous water issues, without any clear underlying philosophy linking or binding them. There were also several sessions on very similar topics. Due to the vast choices of the sessions organized, participants mostly went to the ones that interested them, and where their individual ideologies and views were most likely to be supported. A good example was the many sessions on dam-related issues. At one session in the Hague, a speaker passionately claimed that all dam builders should be prosecuted through the war crimes tribunal since building of dams is a crime against all humanity! At another session, a different speaker suggested that dams are absolutely essential for poverty alleviation in the developing world, and thus many more dams must be built. Although the two statements were diametrically opposite, no one challenged such statements. Thus, both the speakers and the audience present at the two sessions probably returned to their homes thinking that the Forum participants had basically accepted their views.

Accordingly, a major constraint of the second and the third Forums was the inadequacy of any sustained attempt to link the various sessions on similar topics, or related topics. Furthermore, the papers and discussions of the Hague and the Japan Forums are now irretrievably lost, since no effort was made to collect, synthesize or disseminate them. What is available is a set of somewhat general and superficial summaries which are not of much use to any serious water professional or institution. Regrettably the problems were very similar for the Fourth Forum in Mexico City as well.

The situation was better in one way at Japan, compared to the Hague. If the dam issue is considered, at Kyoto, a constructive debate on this subject was organized by the International Hydropower Association and the International Rivers Network. The views on large dams of these two institutions are diametrically opposite. The two groups listened to each other, and there was the beginning of a dialogue between the opposing camps. The debate could not have changed the views of the diehards in the two camps, but it may have had some impacts on some members of the audience, who were somewhat neutral and open-minded on the issue. While this debate was a plus for the third Forum, on the issues of dams, the sheer number of sessions held at Kyoto ensured that very few participants, if any, had an integrative view of the relevant discussions on the dam-related issues. In addition, the overall views on the needs for and the impacts of large dams at different sessions were different, sometimes totally different. These views were often ideological and not based on facts or objective analyses.

All the three World Water Forums had another major shortcoming. Not even one seriously discussed or raised the water issues of the future, say, beyond the post-2010 period, let alone to 2020 and beyond. All the three Forums consistently argued that "business as usual is not an option," but then behaved as if it was the only option available. The world is changing rapidly, and real visionaries are urgently needed to develop future water visions of the world. All the visions presented at the Hague, at national and regional levels, as well as sectorally, were far too general for any possible practical use. Thus, and not surprisingly, in Japan, all these visions simply disappeared from the Forum agenda. Accordingly, continuity and interlinkages between the second and third Forums left much to be desired. In fact, there have been no real interlinkages between the four Forums held thus far. For all practical purposes, all the four Forums held thus far have been individual and discrete events, with no real interlinkages, or continuing discussions on priority issues, where the results or the conclusions from one event was taken and followed through in the next. This shortcoming is one of the main causes which has ensured very low impacts of these four megaconferences.

A significant percentage of the sessions at the Third Forum was similar to those in the Hague, in terms of topics, overall poor quality of the presentations and absence of quality control. The discussions were exclusively past and present oriented. No new innovative idea came out from the three World Water Forums, no new ground was broken, and no new commitments were made by the governments present in terms of new investments, or water-related activities that would not have happened without these events.

It is interesting to note that the Dutch Minister for Development Cooperation, Agnes van Ardenne, said categorically at Kyoto that large-scale megaconferences like the World Water Forums have no future. The view of the Dutch Government is noteworthy since it hosted and financed the Second World Water Forum. Her views, and the reasons thereof, should have been carefully considered by the water profession in general, and the World Water Council in particular, prior to the hosting of the subsequent forum in Mexico City.

The questions that must be asked at present are as follows: Are these megaconferences worth their costs and the efforts needed to organize them, especially in terms of their eventual and overall impacts? Are there better and more cost-effective alternatives, where the world can get "bigger bangs for smaller bucks"? Unfortunately, these types of questions are not even being asked at present, let alone being answered. Everything considered, the time has come to stop being politically correct and claim everything is fine with these megaconferences. Past performances should be objectively evaluated in order to develop a cost-effective and a high impact road map for the future.

It is thus essential for the water profession to critically and objectively assess the overall impacts of the past water-related global megaconferences. This must not be a pseudo-evaluation, carried out with rose-coloured glasses and by the people who have been directly associated with the organization of these events, as have mostly been the case in the past. Nor should the results of these evaluations be kept confidential: otherwise future progress can at best be incremental, and the overall governance process will neither improve significantly nor be transparent.

The evaluations must be independent, objective, comprehensive and usable. The results of such evaluations should be used to define what other alternatives may be available in the future to obtain significantly better results and impacts, but in a more cost-effective and timely manner.

Because of the absence of any reliable assessment of the global water-related megaconferences of the past thus far, the Third World Centre for Water Management, with the support of the Sasakawa Peace Foundation of the United States, initiated a project which objectively and realistically examined the past events in order to identify their contributions to ensure a water-secure world for the future. This was not a pseudo-evaluation, as were carried out for some of the past events, but a serious attempt to evaluate objectively the impacts of the past megaconferences, and also try to see if there are better and more cost-effective alternatives. The results of this evaluation are outlined in the book. It clearly shows that all is not well with the megaconferences of the water world.

1.9 Concluding Remarks

The water management profession is now facing a problem, the magnitude, complexity and importance of which no earlier generation has had to face. In the early part of the 21st century, our profession really had two choices: to carry on as before with a "business as usual" attitude that attempts to solve future complex problems on the basis of experiences from simpler problems of the past, or consider in earnest an accelerated and truly genuine effort to identify the real problems of the future and face the overwhelming challenges collectively and squarely by implementing workable "business unusual" solutions within the short timeframe available to us. One of the main lessons of the past has to be that the time for rhetoric and using one minute sound-bites is now over. We must develop urgently new and cogent paradigms and solutions which can be operationalized in developing countries and in the fields. Conceptual attractiveness alone is no longer adequate. Activity can no longer be considered to be equivalent to progress.

Globally, water is likely to become an increasingly critical resource issue for at least the next two decades in the developing world. Equally, forces of globalization, urbanization, population growth, food, energy and environmental securities, technological developments and information and communication revolution are changing the planning and management requirements of the water sector with stunning speed. The world is moving into a new kind of economy as well as to a new kind of society, where new mind-sets and knowledge are needed to resolve increasingly complex and interrelated issues. The water sector is no exception to this development. Whether we like it or not, the world of water management is likely to change more during the next 20 years compared to the past 2000 years. The past experiences will often provide no guidance, or at best only limited guidance, during this period of explosive change and increasing complexities, uncertainties and unexpected turbulences. The stakes are high, but equally they give us new opportunities to improve water management practices very significantly like never before in human history. These complex trends and changes have to be identified and successfully managed. The opportunities are clearly there, and the tasks are doable. Accordingly, the water profession as a whole must rise to meet these challenges successfully and in a timely manner. These are priority issues that future water-related megaconferences must address firmly and squarely.

In terms of megaconferences, one very fundamental question that needs to be asked and answered is how can these events be carefully planned, organized and structured so that they can make meaningful contributions to ensure a water-secure world of the future. The present effort to assess the outputs and actual impacts of the megaconferences is a step in this direction.

References

Ait Kadi M, Shady A, Szollosi-Nagy A (eds) (1997) Water, The World's Common Heritage: Proceedings of the First World Water Forum. Elsevier Science, Oxford

Asian Development Bank (2007) Asian Water Development Outlook 2007. Asian Development Bank, Manila

Biswas AK (1988) United Nations Water Conference: Implementation Over the Past Decade. International Journal of Water Resources Development 4(3): 148–159

Biswas AK (1993) Water Missing from the Agenda. Stockholm Water Front 2: 12–13

Biswas AK (1997) From Mar del Plata to Marrakech: From Rhetoric to Reality. In: Ait- Kadi M, Shady A and Szollosi-Nagy A (eds) Water, the World's Common Heritage. Elsevier Science, Oxford, pp. 27–35

Biswas AK (2000) Water for Urban Areas of the Developing World in the Twenty-First Century. In: Uitto JI, Biswas AK (eds) Water for Urban Areas. United Nations University Press, Tokyo, pp. 1–23

Biswas AK (2006) Challenging Prevailing Wisdoms: 2006 Stockholm Water Prize Lecture. Third World Centre for Water Management, Atizapan, 10p.

Biswas AK (ed) (1978) United Nations Water Conference: Summary and Main Documents. Pergamon Press, Oxford

Biswas AK (ed) (1978a) Water Development and Management. Proceedings of the United Nations Water Conference, 4 Volumes. Pergamon Press, Oxford

Biswas AK, Tortajada C, Izquierdo-Aviñó R (eds) (2009) Water Management Beyond 2020, Springer, Berlin

Mageed YA (1978) Opening Statement. In: United Nations Conference: Summary and Main Documents. Asit K. Biswas (ed), Pergamon Press, Oxford

Mageed YA (1982) The United Nations Water Conference: The Scramble for Resolutions and the Implementation Gap. Mazingira 6(1): 4–13

Rodda J (1993) Water Moving up the Agenda. Stockholm Water Front, No. 3: 6

Tortajada C (2006) Water Management in Singapore. International Journal of Water Resources Development 22(2): 227–240

World Commission for Water in the 21st Century (2000) A Water Secure World: Vision for Water, Life and the Environment. World Water Council, Marseilles

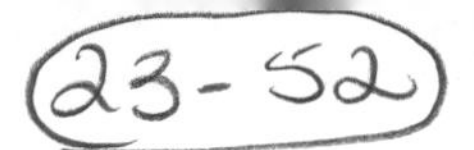

2 Mega United Nations Conferences: Help or Hindrance?

Morris Miller

2.1 Approach

The approach of this chapter is, first, to assess the effectiveness of the policy of the international community that currently places great reliance on the staging of large-scale broad-scope conference under the auspices of the United Nations to address critical issues of global concern such as those related to development, poverty, energy, environment, water, etc. Two United Nations conferences are examined as case studies with regard to both process and substance: the recent Millennium Development Conference and the 1981 United Nations Conference on New and Renewable Sources of Energy. Second, the focus is placed on identifying the type of institutional obstacles that need to be surmounted if the objectives of these conferences are to be attained and on why the traditional mega United Nations conference is more likely to be a hindrance rather than a help in overcoming these obstacles. The third part is devoted to drawing lessons and putting forward proposals that could get around these obstacles with the United Nations playing a different more focused role to enhance the possibility and probability of achieving the desirable and ambitious goals of the United Nations megaconferences.

2.2 Introduction

> Words! Words! Words! I'm so sick of words!
> I get words all day through, first from him, now from you!
> Is that all you blighters can do?...
> Sing me no song! Read me no rhyme!
> Don't waste my time, show me!
> Make me no undying vow. Show me now!
> —Alan Jay Lerner's lyrics from *My Fair Lady*

Today, more than half a century after President Roosevelt and Prime Minister Churchill articulated in the Atlantic Charter the challenging goal of a post-war world free of fear and free of want, there are still more than 2 billion people or one-third of humankind mired in dire poverty with all that implies in terms of deprivation not only of material goods and services, but also of hope. One of the principal means that the international community has adopted to address this challenge is to stage a series of conferences under United Nations auspices. The list of conferences over the last four decades is long but 1970 might usefully be cited as the beginning of a cycle related to energy, water and other natural resources and all of them to development and to the environment (see Box 1).

Box 1

The list of the relevant large-scale broad-scope conferences launched under the auspices of the United Nation since 1970 is as follows: United Nations Conference on Environment and Development (1972), United Nations Conference on Science and Technology (1980), United Nations Conference on New and Renewable Sources of Energy (1981), World Summit for Children (1990), World Conference on Education for All (1990), 2nd United Nations Conference on the Least Developed Countries (1990), United Nations Conference on Environment and Development (1992), International Conference on the Least Developed Countries (1992), International Conference on Nutrition (1992), World Conference on Human Rights (1993), International Conference on Population and Development (1994), Global Conference on the Sustainable Development of Small Island Developing States (1994), World Conference on Natural Disaster Reduction (1994), World Summit for Social Development (1995), 4th World Conference on Women (1995), 2nd World Conference on Human Settlements (1996),World Food Summit (1996), Millennium Summit (2000), 3rd United Nations Conference on the Least Developed Countries (2002), International Conference on Financing for Development (2002), World Food Summit: Five Years Later (2002), World Summit on Sustainable Development (2002). A summit of world leaders convening in Monterrey in 2002 pledged financial support for the achievement of the Millennium Development Goals (MDGs).

In addition there have been international meetings called World Water Fora that focused on the issue of water management: Marrakech 1997, the Hague in 2000, and Kyoto 2003.

Despite progress in terms of morbidity rates, longevity and illiteracy and other indicators, there still remains a deplorable state of affairs of deep poverty. This state of affairs prompts the question as to whether the United Nations' many global megaconferences have made much of a contribution in the struggle to reduce the world's severe deprivations; and if it has not done so to any significant degree, to ask why and what are the alternative means to attain the desired goals?

It has been claimed repeatedly that these conferences have been helpful in many ways and, in this regard, it might suffice to quote one commentator (Taylor 2003: 157) who has made that point in a succinct manner in stating that the conferences have been:

> A focus of heroic effort by non-governmental organizations throughout the world, ...(that have) promoted intense interaction between members of participating governments; ...(that have) added something new to multilateral diplomacy (in) identifying a core of

> agreed values and purposes which formed the basis of special actions and programmes over a very wide range of human interests and needs; ... (that have) strengthened among diverse groups a sense of common destiny, and (that have set forth) a global agenda ... (that) frequently included specific targets, timings and policy proposals.

Above all, he concludes, ultimately the most important outcome of the conference process has been "entrenching of multilateralism", by which he means that by virtue of the conferencing process "the United Nations System has become a forum of obligation."

It is fair to say that there is a widely held contrarian view that maintains that this is not a realistic assessment, that the inducements or pressures contained in the conference resolutions have been neither specific enough nor strong enough to assure the desired follow-through "on the ground" at either the level of global or national governance. There, indeed, may be as an outcome of these conferences a greater sharing of "a sense of common destiny" and a sense of "obligation" but these are ambiguous concepts operationally and, therefore, not helpful enough to lead to *concrete* measures as differentiated from a form of action that is basically preparing more meetings and writing more reports. Given the talent and the money and the hopes that have gone into preparing and running these megaconferences, the conferences might even be deemed to have been a drag rather than a stimulus to action.

A sceptical view has been put forward by *The Economist.* In its 9 September 2004 issue, in an article headed, "The United Nations has set benchmarks for progress in poor countries—are they (of) any use?", a cryptic answer is provided by another heading: "ends without means". The following assessment is offered:

> The weakness of the whole Millennium Development Goals (MDG) concept is that it wills the ends without willing the means—something which the United Nations, perforce, has come to specialize in.... It remains questionable whether the MDG exercise with its unimpeachably good intentions and its proliferating bureaucratic overhead, has done any good at all, on balance.... In fact, how far the MDG initiative is making a difference, one way or another, is unclear.
>
> The United Nations observes that 'many countries are in the process of retooling development programmes and strategies in line with the MDGs'. How odd: were those governments hitherto unconcerned about poverty or AIDS?

Parodying Marie Antoinette, this is followed with the advice: "let them eat reports". There has never been a shortage of reports as the follow-up "activity".

The positive impact "on the ground" will likely remain inadequate in terms of the amount of aid and the use to which it is put. The article in *The Economist* poses a question on this vital issue: "has the MDG process (the series of megaconference culminating in the MDG) at least succeeded in directing more aid to the right uses?" And their answer is: "not really". As evidence of this, they note that the total amount of official development aid flows (ODA) for the poorest countries is \$68.4 billion or only 0.25% of the donor countries' aggregate annual incomes, an amount that is still far short of the goal of 0.7% of national income that was agreed upon at a United Nations conference several decades ago. But what is more damning is the fact the pledges made at the recent Monterrey Summit, where the world's financial ministers convened post-Millennium Development Conference, would only increase ODA by an amount that would raise the percentage to only 0.3% by 2006. This is less than half the target that was pledged a quarter century ago!

There are others who share this scepticism as to whether the mega conferencing process has succeeded in increasing aid flows, directing more of the aid to the right uses and improving the effectiveness of these aid flows. Focusing on the water-related aspect of such conferences, Asit Biswas, the president of the Third World Centre for Water Management, has tackled the issue succinctly and clearly. In an article titled, "From Mar del Plata to Kyoto: an analysis of global water policy dialogue" (Biswas 2004: 87) he has this to say about past conferences:

> The question that must be asked at present is, are these mega-meetings worth their costs and the efforts needed to organize them, especially when their final and overall impacts are considered.... It is high-time that we stop being politically correct and objectively review our past performances in order to develop a cost-effective and impact-oriented road map for the future.... Conceptual attractiveness alone is no longer adequate.

He goes on to quote the Dutch Minister for Development Cooperation, Agnes van Ardenne who "said categorically at Kyoto that large-scale conferences like the World Water Forum have no future" and to also quote the head of a United Nations agency who remarked sarcastically: "all our delegates are honourable, all our backgrounds documents are excellent, and all our meetings are outstandingly successful."

Even those who attempt to give a favourable assessment are ambiguous in terms of the accomplishments of these megaconferences as, for example, a United Nations official, Masumi Ono, who, writing on the issue of the follow-up to United Nations conferences, gave this qualified assessment: (Ono 2001: 180)

> The United Nations through its series of global conferences has contributed to building consensus and norms by initiating a continuous process of mobilizing political will…it is too early to say anything definitive about the success or failure of many of the items on the agendas of United Nations conferences.... The challenge now is to operationalize the consensus and norms in a comprehensive and coherent way.

Would it, therefore, be impertinent to pose the following question: if, as Ms Ono puts it, *now* is the time to operationalize the resolutions and agreements about goals and procedures, what have all the blighters at these conferences been doing all this time?

George Bernard Shaw's heroine of the play *Pygmalion*, Liza Doolittle, comes to mind: she became justifiably exasperated with "her betters" in her struggle to learn how to act and speak like "a fair lady". Unlike Pygmalion, the mythological king of Cyprus, who with a stroke brought to life a statue he had made of the goddess, Aphrodite, with whom he had fallen in love, there are no magic wands, nor magic words that can turn a vision into reality. After the talk of what needs to be done, it is the follow-up in terms of making institutional and policy changes that is the hard slog up a steep slope full of obstacles. Only in mythology can there be the attainment of wishes without Herculean effort over a considerable span of time.

One of the key elements of this Herculean effort is political will and this is reflected in the financial muscle applied to the task. There is first of all the issue of obtaining sufficient public support to cover the necessary financial costs of proposed policies with their ambitious programmes and projects. This is a formidable challenge in the light of the fickle nature of gaining public support for the proposals emerging from these conferences. There is the basic matter of whether the process of conferencing with its attendant publicity succeeds in spreading knowledge and enhancing trust in the desirability and feasibility of the proposals. To illustrate how feeble this follow-through impact happens to be, we could cite the reaction of Canadians: it was reported in an issue of *The Ottawa Citizen* (2 October 2004) that a poll taken on 5 September 2004, of the opinion of a representative sample of Canadians, revealed that four out of five persons supported ratification of the Kyoto Protocol, but nearly two out of three had no idea what it was about, and half had never heard of it! The journalist reporting this cites a study that found that the annual cost of Kyoto is estimated to be about C$4,700 per Canadian for the next 5 years, an amount that is roughly equivalent to *per capita* spending on health care and then the journalist goes on to observe that when this fact was mentioned to those being questioned, the pollsters found that "support shrinks as people understand its (financial) impact".

There is, comparatively speaking, a much lower cost in financial terms for preparing, organizing and staging the event and some follow-up business such as preparing reports and publicizing them. This is a cost that is borne by the delegates, by the United Nations itself and by the host country (when the event takes place outside United Nations facilities). This expenditure is rationalized as a democratic way of policy-making by informing the public and assuring greater public support for the proposals that emerge. It is generally accepted without challenge given that these are events that have taken on the nature of traditional activities of a global institution such as the United Nations System, and, in any case, their financing is trivial in amount for the world community and, in any case, are seen – especially by non-governmental organizations or what is called, "civil society" – as appropriate political responses to troubling issues of global scope.

But this cost in financial terms pales in significance beside two other costs of a non-financial nature:

- One is the "opportunity cost", that is, the factor of delaying what might otherwise have been done with that time, talent and money that could possibly have been spent more effectively than indulging in talk, talk, talk and promises, promises, promises.
- Then, in addition, there is the collective psychic cost incurred when expectations are aroused and disappointing outcomes become apparent after a lapse of time. What follows is greater scepticism and its close cousin, cynicism and, from that, diminished trust in the political leaders who made the decision to take the conference talking/promises route. This, in turn, would undoubtedly make it more difficult to find the financial and other forms of support for corrective action that goes beyond posturing and rhetoric.

But, in the final analysis, the over-arching questions that need to be answered are as follows:

- Have the United Nations megaconferences provided the launching pad for effective initiatives that would involve changes in institutions, policies, projects and practices that are commensurate with the nature and scope of the global-wide challenge in terms of poverty alleviation, reducing inequality and instability, protecting and expanding human rights, assuring access to educational and health services (and, in that connection, potable water), etc.? (see Box 2).
- Given their nature (size, complexity, etc.), could they ever have succeeded in doing so, and, if not, what are the alternative approaches that hold promise as an initiative at the international level of governance?

Box 2

Water has been a key deprivation with over 1 billion people unable to access clean drinking water, more than twice that number lacking access to adequate sanitation facilities and about four times that number without access to sanitary wastewater disposal facilities. (The World Health Organization defines "reasonable access" as the availability of 20 litres per person per day from a source within 1 kilometre of the user's home.) An estimated 12 million people die annually from the scarcity of clean water leading to waterborne diseases. Yet, on the present trajectory according to a recent report, The World Water Report (UNESCO 2003), by 2050 there will be severe water shortages confronting 7 billion people, that is, about two-thirds of the projected global population. At the same time, it bears noting that the United Nations Committee on Economic, Social and Cultural Rights explicitly recognizes the rights to water as a "human right", that the Director-General of IFPRI has stated in the foreword to one of his organization's publications that "the defining issue of the 21st century may well be the control of water resources" (Rosegrant 1997), and that others have put forward the case that "the nexus between development, water, and human rights is well established in the international legal regime" (McIherney-Langford and Salman 2004).

One way to proceed is to examine some types of conferencing that by their very nature illustrate how not to proceed and, by inference, illuminate a way to proceed. One such conference of the how-not-to-proceed variety is the broad-based Millennium Development Conference that took place at the very beginning of this millennium. It has set out an ambitious and eminently desirable array of goals as a stimulus to action and as a guide to institutional and operational initiatives that should and could be undertaken at various levels of governance.

Another type of conference that merits examination is one that has focused on a particular aspect of the global *problematique*, as, for example, the United Nations Conference on New and Renewable Sources of Energy (UNCNRSE) that took place in Nairobi in 1981 a quarter of a century ago, and, for our purposes, has the virtue of being related to the water management theme of this workshop and to the many facets of life in which energy, the environment and water issues are intimately connected.[1]

Though both these conferences are representative of two types of approach, the broad and the sectoral, they are characterized by similarities in terms of scale, organization and procedures that are typical of United Nations conferences. The assessment of their success or failure and the lessons to be learnt from a broad-brush description and analysis of both of them should have implications about United Nations conferencing in general as an approach to tackling global-scale problems and about possible alternative approaches.

2.3 Two Case Studies: The Millennium Development Conference and the United Nations Conference on New and Renewable Sources of Energy

2.3.1 The Millennium Development Conference

> There'll be crumpets and tea without you.
> Art and music will thrive without you.
> Somehow Keats will survive without you.
> And there still will be rain on that plain down in Spain,
> even that will remain without you.
> I can do without you.
> — lyrics from *My Fair Lady*

There have been several broad-based United Nations conferences on such themes as human rights, social development and, most recently, to mark the new millennium, a catch-all one called The Millennium Development Conference from which emerged The Millennium Declaration with its long list of Millennium Development Goals (MDGs). It set targets for reducing key deprivations such as access to educational and health facilities and such basic needs as clean drinking water and sanitation. A perusal of the eight goals, as listed in Box 3, shows that four MDGs are concerned with health and one of the four is focused on the aspect of access to safe drinking water and sanitation.

1 And it has the additional bonus of being a conference with which the author was intimately involved as its Assistant Secretary-General and which, therefore, could provide an insight from the special vantage of an insider and could enable lessons to be learnt on the basis of a retrospective vision of more than two decades.

Box 3

The MDGs to be achieved by 2015 are the following:

1. reduce by half the proportion of people living on less than $1 per day and those suffering from hunger;
2. achieve universal primary education;
3. eliminate gender disparity in primary and secondary education;
4. reduce by two-thirds the mortality rate among children under five;
5. reduce by three-quarters the maternal mortality rate;
6. halt and begin to reverse the spread of HIV/AIDS and the incidence of malaria and other major diseases;
7. ensure the environment (including reduction by half those without access to safe drinking water and sanitation facilities) and significantly improve the quality of life of at least 100 million slum dwellers by 2020);
8. develop a global partnership for development.

As the United Nations Secretary-General observed, "at this stage of global development, such deprivations were deemed to have persisted for too long and constituted an affront to the conscience in what purports to be civilization. This goal-setting exercise was followed by more meetings in Doha, Monterrey and in the summer of 2002, in Johannesburg, the World Summit on Sustainable Development (WSSD) attracted 40,000 persons: 1200 heads of state and government officials, executives of 500 corporations and thousands more from non-governmental organizations. There had been backsliding since the Conference on the Environment and Development held in Rio in 1992; the declared challenge was to get on track to achieve "sustainable development", an over-arching goal that has become a cliché and, as such, is regrettably vacuous in operational terms (Miller 2005).

There has long been recognition of the need for these meetings to get beyond the traditional communiqués that have been comprised of wish-lists of goals or targets with timelines that are devoid or weak on the aspect of implementation, that is, how such matters as institutional changes and financing and other requirements are to be achieved. Thus, the mandate of the General Assembly for the Johannesburg Summit called for going beyond an assessment of what had been and what had not been accomplished since the Rio Conference. The Johannesburg meeting provided an occasion to revise the goals and to harden up the soft language of Rio with regard to how to attain the goals, including the challenging aspect of finding the necessary financing that in 1992 at the Rio Conference was estimated to be over $600 billion of which $125 billion would be expected as the contribution of the international community. A Plan of Implementation was drafted along with a firm financial commitment of only $3 billion replenishment for existing programmes and with an expectation of raising the necessary additional funding through a programme called Type II Partnerships that would be voluntary agreements with the private sector for financing specific initiatives.

The Millennium Conference of 2000 went further: it established an entity called the Millennium Project, the mandate of which is to advise the Secretary-General. The Administrator of the United Nations Development Programme (UNDP) would be assigned to act as chairperson of a special unit called the United Nations Development Group (UNDG). Coordinators working under the leadership of Professor Jeffrey Sachs have been assigned to head 10 Task Forces comprised of world-class scholars, personnel of United Nations agencies, public and private sector institutions, and non-governmental organizations. Each task force has been charged with producing a report focused on assigned themes that are wide ranging (see Box 4). The mandate of the Millennium Project is clear and ambitious – to recommend by June 2005:

> operational strategies for meeting the Millennium Development Goals (MDGs) (that) includes reviewing current innovative practices, prioritizing policy reforms, identifying frameworks for policy implementation and evaluating financing options, the ultimate objective (of which) is to help ensure that all developing countries meet the MDGs.

"So", wrote one commentator, "the summit was anything but a complete failure" (Ruffing 2002: 40). To which the appropriate reply might be "let's wait and see."

Box 4

The themes of these task forces are the following: (1) poverty and economic development, (2) hunger, (3) education and gender equality, (4) child and maternal health, (5) major diseases (HIV/AIDS, malaria, TB and others) and access to essential medicines, (6) environmental sustainability, (7) water and sanitation, (8) improving the lives of slum dwellers, (9) open, rule-based trading systems, and (10) science, technology and innovation. Further details on these reports can be found at www.unmilleniumproject.org.

The interim reports of the 10 Task Forces have already been made available. On the face of it, to judge by the quality of the analysis and the evident concern of the authors for the factor of feasibility, there would appear to be some promise that actions commensurate with the challenge might actually follow. Scepticism, however, arises from the gap between promise and performance revealed by a study of the historic record of past United Nations megaconferences, especially those of ambitiously broad scope. Short of some means of enforcement, reliance has traditionally been placed on the weak reed of moral suasion underpinned by regulation, offering incentives to do "the right thing", and by monitoring followed by publicity that might shame governments who fall short of commitments, if they even have made commitments by signing on a dotted line.

This line of reasoning has given rise to the World Bank's publication of *The Global Monitoring Report 2004, Overview: From Vision to Action* (World Bank 2004a). On the opening page the authors are quite explicit in stating that:

> the themes of implementation and accountability constitute the fundamental motivation behind the global monitoring initiative… With broad agreement on the goals and strategies to achieve the MDGs, *the task now is implementation*. (emphasis added)

Now implementation? Does this suggest that not much of significance has been achieved at these meetings beyond identifying a problem and setting goals? The usual answer is that goal-setting is a spur to action.

To assess this answer there is a need to examine the record with regard to the implementation follow-up. While *The Global Monitoring Report 2004* and *Rising to the Challenges: The Millennium Development Goals* (World Bank 2004b) lists progress towards attaining some of the MDGs it reports, as well, on the lack of progress and, in that regard, it states that only about one out of six of the developing countries are currently on track to reach the relevant MDGs and that this shortfall still leaves more than 10 million children in their countries dying before reaching their fifth birthday and as many as 500,000 women dying during pregnancy or in the process of childbirth. On the positive side, there is a list of achievements that include attainment of some of the MDGs well before the target date and of progress towards achieving them. The progress appears impressive:

> globally, adult illiteracy fell by half over the past 30 years, while life expectancy at birth rose by 20% over the past 40 years… from 1990 to 2002 Vietnam reduced poverty from 51 to 14%… over the course of the last 15 years Botswana doubled the proportion of children in primary school… in the 1990s Benin increased its primary enrolment rate and Mali its primary completion rate by more than 20%… between 1990 and 1996, Mauritania increased the ratio of girls to boys at school from 67 to 93%…

What is noteworthy is that this progress has been made without any contribution of the 10 Task Force Reports of the Millennium Project. It is on the basis of the proposals contained in these reports with regard to programmes and projects and related institutional changes that governments are being asked to take action. The point of this is simply to illustrate that there is no evidence of a necessary relationship between the conference process, including its follow-up in terms of report writing, and the implementation phase. There are too many factors at play to correlate the desired actions on the ground with the articulation of goals and of proposals emanating from these conferences as a cause-and-effect relationship. Indeed, as the authors of the cited World Bank reports point out, rapid progress is possible given "good policies and the support of partners".

Perhaps the most cogently expressed put-down of such conferences is that presented by Stephen Rosenfeld, in an op-ed piece in the 16 September 1994 issue of *The Washington Post* titled, "The Cairo Mandate" when he posed the following questions in connection with the International Conference on Population and Development that was held in Cairo in 1994:

> What could a conference like this actually do? Did the world really need a population conference to determine that the most rewarding remedial actions is to spend more on such programmes as the education of girls and the empowerment of women?"

He noted that "(these questions are) asked somewhat dismissively, or despairingly", and went on to make an observation about the Cairo Conference that would seem to apply to all other United Nations megaconferences:

> Cairo's accomplishment lies in the essential chemistry of social change: converting a slowly won new expert consensus on population and development into a virtually worldwide political consensus, thus fortifying advocates returning to battle in their own countries on an agenda that assigns new weight to a developmental approach to women's rights and health...You are left with the largest calculated act of social engineering in history.

Is it any wonder that he is prompted to characterize the United Nations in staging these conferences as a "dumping ground of desperate hopes... (that) badly needs a shot of relevance and effectiveness"?

This gives rise to the thought that these large-scale well-publicized United Nations conferences were very likely conceived by politicians who are under pressure to address serious problems or crises that call for action at the level of international governance: these conferences are a showcase that demonstrates their shared concern with their constituents – *and is a form of action*. Whether it is useful in the sense of giving rise to follow-up action "on the ground" is a secondary consideration for these politicians. At best, through the staging of such conferences, they could hope to achieve the modest goal of raising public awareness and spreading some key elements of knowledge about such issues while not being obliged to do much beyond talking the good talk and signing operationally vacuous do-good agreements. The mode of operation of these conferences lends itself to this viciousness since the most reluctant participant among the donor industrialized nations can impose their will with regard to financial or other commitments. The old adage applies: the convoy cannot go faster than the slowest ship. And what is worse, the slowest ship may be deliberately stalled in the water or sail off course when mixed signals are received as to the direction and speed to be taken.

In this regard the observation of a Canadian participant at the conference that gave rise to Agenda 21 seems apropos. Peter Padbury, a former Director of the Ottawa-based Canadian Council for International Cooperation, in a report entitled, *UNCED and the Globalization of Civil Society: Lessons for United Nations Reform* (Padbury 1993: 4) had pertinent comments on the issue of the negotiations at that conference, a process that is all too typical of broad-scope megaconferences:

> as negotiations proceeded the wording became more general and the commitments less precise...The negotiations did not seem connected to problems or actors in the real world...The principle strategy of governments seemed to be to ensure that nothing happened that obligated their governments to make any changes...

The debate on financial resources to pay for Agenda 21 was illustrative. A great deal of time was spent on the finance question (but) the discussion on finances was a stalemate that was never formally resolved.

He went on to characterize this as,

> an amazing process, but, like many United Nations conferences, it suffered from a number of problems (such as putting) emphasis on negotiations rather than on an effective change process, on sectoral rather than on system level change, on the nation state rather than on the planetary system...

The mixed signal metaphor would seem to be suggestive of what is likely to transpire when the attribute of interconnectedness arises with its attendant complexity for both analysis and its operational implications. Agenda 21 exemplifies this phenomenon that makes it very difficult to arrive at decisions as to what needs to be done and how it is to be done in terms of executing agencies and financing and other aspects. This difficulty is also clearly exemplified in the case of the World Summit for Social Development at Copenhagen in March 1995 that, like the Millennium Development Conference, was very broad-based. The former United Nations Secretary-General, Boutros Boutros-Ghali, made a pertinent observation (Boutros-Ghali 1999: 171) that recognizes the interconnectedness of the broad-based and sectoral conferences.

> The ills that societies feel most acutely all have social origins and social consequences (necessitating) focus on the urgent and universal need to eradicate poverty, expand productive employment and enhance social integration... (By placing the) focus entirely on the most deprived segment of global society the World Summit for Social Development at Copenhagen in March 1995 stressed the interconnectedness of the entire continuum of United Nations conferences.

The stress arising from this interconnectedness of the entire continuum of United Nations conferences appear to have been too great. Even the eminently desirable and seemingly feasible 20/20 proposal (that would have obliged the international community to devote 20% of their aid to social programmes and projects with the recipient developing countries in their turn committing to devote 20% of their budgets to the same type of programmes and projects) could not emerge as a recommendation from the deliberations of the United Nations' Social Summit in Copenhagen. Other concrete recommendations failed to make it to the ultimate draft; only very vague statements survived the weaning-out process. As Professor Michael Schecter observed in his conclusion to a book on the impact and follow-up of United Nations-sponsored world conferences (Schecter 2001: 219),

> One of the truths of such conferences is that they cannot paper over significant differences amongst governments and cultures on salient and contentious issues, especially when they are negotiated at a widely publicized global conference... The United Nations-sponsored conferences...seek to focus on one major issue-area whereas globalization makes that difficult, if not actually infeasible or dysfunctional.

But, as we shall see in the last section of this chapter, this does not mean abandonment of a facilitating role for the United Nations via smaller and sharply focused conferencing where meaningful outcomes at the international level of governance are possible, if not likely. The United Nations Conference on New and Renewable Sources of Energy would seem to be a good candidate to illustrate this point: despite being focused, it was not small and it failed to have any significant impact. Though it is, so to speak, "a vintage conference" having taken place in 1981 after, and in response to, the so-called "international energy crisis" of the 1970s, the lessons to be learnt apropos the effectiveness or non-effectiveness of United Nations megaconferences are pertinent to the challenge we face today.

2.3.2 The United Nations Conference on New and Renewable Sources of Energy

> Just you wait, 'enry 'iggins, just your wait!
> You'll be sorry, your tears will be too late!
> You'll be broke and I'll have money;
> Will I help you? Don't be funny!
> Just you wait, 'enry 'iggins, just you wait!
> — lyrics from *My Fair Lady*

How long do we have to wait? More than a quarter of a century has passed since the so-called "international energy crisis" was precipitated by OPEC's tripling of oil prices and the world was gripped with fear at the prospect of living in the dark. The neo-Malthusian Club of Rome's publication, *The Limits to Growth*, became a best-selling tract rivalling the sales of the Bible. The financial/economic impact of a three-fold rise in oil prices and the dire implications for the global economy, and particularly poor oil-importing countries, prompted the United Nations members to convene a major conference, the United Nations Conference on New and Renewable Sources of Energy that, after 2 years of preparation, was held in Nairobi in 1981. To what effect?

Here was a conference arising out of an urgent need to address an abrupt large increase in the price of an essential commodity. The "oil shock" of the 1970s brought on a realization that there would need to be a search for alternatives to fossil fuels in part to assure an energy supply and in part to reduce environmental damage. The Nairobi Conference was, thus, born out of recognition of the need for research and other measures to bring forth alternative energy sources in a form and at a price that could make a significant global impact. The conference's mandate was to identify the problems and recommend steps to meet them through national and international-scale policies and programmes.

After more than 2 years of preparation, a so-called Plan of Action emerged, a distillation derived from the discussions in over twenty workshops and from the many background papers prepared by the workshop participants, special consultants, NGOs and government agencies. The recommendations pertained to national and internationally sponsored actions, including new institutional arrangements, to promote research and development of alternative sources of energy, particularly those that could displace fuelwood and, therefore, prove of special relevance for developing countries.

If judged by follow-through "on the ground", all this effort and expenditure of talent and money ended in failure. There is little residue in the public consciousness of its deliberations and of its proposals for action. It is rare to even find in the media any reference to the Nairobi Conference when the major environment-related conferences are listed and discussed. We hear of Stockholm, Rio, Kyoto and Johannesburg but the one that focused on environmentally benign energy, the key to a better environmental future, is almost always omitted. But, notwithstanding, one could ask: did the Nairobi Conference's Plan of Action have some impact on the conferences that followed? It would be hard put for anyone to demonstrate that there has been any follow-up worth mentioning to the initiatives that were put forward in the Nairobi Conference or in any of the subsequent United Nations conferences on the same theme.

It is relevant to understand why the Nairobi Conference did not leave an indelible footprint on either the public's consciousness or the politicians. A large part of the reason would seem to lie in the inappropriateness of the process of large-scale conferencing to tackle such international challenges. The oil-importing countries campaigned hard for the launching of the conference, and desperately wanted it to succeed, but were tactically inept as might have been expected when, in the name of G-77 unity, they acquiesced in the appointment of officials from the oil-exporting countries to be their leading spokespersons! They were unprepared for the opposition led by the United States, the Soviet Union and the OPEC spokesmen, all of whom clearly desired that the conference fail. The result is an impressive-sounding Plan of Action but one that was devoid of concrete programme or project initiatives except for those that were known *a priori* to be desirable. What emerged were proposals that were a traditional collection of pious platitudes and exhortations formulated in very general terms that lacked specified follow-up steps regarding institutional arrangements and the financing of actionable programmes (see Box 5). This prompts the question: Was there any point to staging the conference?

Box 5

To illustrate, the list of recommendations were the following:

1. governments should undertake surveys and assessment programmes regarding alternatives to oil, gas, coal and nuclear power;
2. the costs and risks of demonstration of new technologies should be shared so as to accelerate their application;
3. more training should be provided for the personnel in poorer countries;
4. governments should undertake a follow-up study to establish estimates of the funding requirements for the relevant programmes and projects.

This megaconference proved to be an effective way to impede change while, at the same time, pretending to recognize the need for change. It is, unfortunately, a very effective and, therefore, a common tactic of politicians, all of whom (as distinguished from statesmen) operate under constraints that are parochial and short-term. Thus, this approach to addressing a major global challenge does little to overcome the institutional system obstacles to change.

2.4 The Barriers to Effective Action

> Let a woman in your life, and patience hasn't got a chance;
> she will beg you for advice, your reply will be concise,
> she will listen very nicely and then go out
> and do exactly what she wants!!!
> — lyrics from *My Fair Lady*

If the global community is to get what it wants, as identified by the ambitious objectives articulated at the United Nations megaconferences, the nature of the resistance to change has to be understood in its many guises. The barriers to such change can most helpfully be identified under three headings (Miller 1994a):

1. The first is a factor labelled *inertia* with its related attributes such as ignorance, scepticism, fatalism, alienation and such. It is a powerful force that sustains the status quo. If effective action on the requisite scale is to be launched and sustained, the necessary condition is a political base that rests on knowledgeable and committed popular support that is powerful enough to overcome the forces of inertia in all its forms.

The argument has often been made that the United Nations conferences, by virtue of the publicity that they generate, have an impact on government policies by raising public consciousness of the problems and of the possible solutions. It has been argued that non-governmental organizations are the committed and organized voice and, therefore, have been effective in playing a role as "the conscience of the world… by placing issues of social justice on the global agenda" (Schecter 2001: 180). Given the half-life of the publicity – and notwithstanding the valiant efforts of some NGOs, most of whom are pulling in different directions and many of which have no compunction about falsehood and exaggeration in their reliance for public support on the fear factor – this is a rather weak basis for action. The attention of the public drifts quickly to other matters with rationales for inaction in the form of cynicism, scepticism, indifference or ignorance in the form of not knowing or relying on blind faith rather than reason (see Box 6).

2. The second barrier to action is a category of responses to suggested initiatives that can be labelled as *aversion to risk-taking*. This barrier is of very special relevance in undertaking initiatives that are of a significant scale. Thus, people settle for "the second best" even if that fear to embrace significant and swift (and often disruptive) change perpetuates deplorable conditions and also has attendant risks: "Better the devil you know" is the watchword.

This is dramatically evident in the case of a water-related issue, the precarious state of the oceans and fisheries and the understandable short-sighted risk-averse perspective of those who are directly dependent for their livelihood on the oceans and fisheries. The World Summit on Sustainable Development (WSSD) provides the case in point: the WSSD ended with an objective about restoring fish stocks to sustainable levels by 2015 and with suggestions about the measures that would need to be taken to attain this goal within this span of time. The rationale for this item in the Plan of Implementation was that it would hopefully exert pressure on the politicians to eliminate or significantly reduce the subsidies that were contributing to an over-capacity in boats, gear and processing plants. However, no account was taken of the short-term political risks involved in a policy of reducing subsidies when, as a consequence, there was a high probability, if not a certainty, that the suggested course of action be resented and, in any case, would not be followed despite the awareness of the long-term denouement of declining fish stocks and declining employment and income and the slow death of the communities. The politician's signature to a United Nations accord means little in the face of the political risk of the possible or probable loss of the vested interest in maintaining office.

Box 6

The blind-faith factor would appear to be a phenomenon of relevance for policy-makers.

A poll of the beliefs of Americans taken in May 2004 by the Association for Canadian Studies revealed that 7 in 10 believe that hell and the devil exist and that 8 in 10 believe in the existence of a heaven where angels abound. There were more self-declared Satanists than self-declared atheists, 30 to 25%! In the Canadian case, only about 1 in three believes in the devil, over 4 in 10 in the existence of heaven and of hell, but almost 6 in 10 in angels. This religiosity cuts two ways: for some there is hope in perceiving an optimistic denouement if the failings of enough individuals are forgiven and redemption follows, and for others, there is the Apocalypse that is foretold and about which we humans can do little except pray.

Reported in the 21 November 2004 issue of *The Ottawa Citizen*, in an article titled, "In God we (Canadians) don't trust as much as Americans do."

3. This brings us to the third barrier that can be categorized as *vested interests in the status quo*. Resistance is to be expected from those who are fearful that they would likely bear the lion's share of the financial and other costs of change without assurance of commensurate benefits or pay-off. There is no institutional or policy mechanism for winners compensating the losers associated with change and the reason is simply greed: those who have it, want to keep it, and want to get more of it! And the way to do that is to resist significant changes to the prevailing systemic arrangements.

These institutional barriers are especially operative with respect to finding enough funds and talent to adequately support programmes and projects under United Nations auspices that are designed for the "global common good". The factor of financing is especially applicable to those activities when there is asymmetry with respect to both the costs (who bears them?) and the benefits (who reaps them, when, and how much?). Given the magnitude of the capital required to meet the ambitious MDGs, the financial aspect of vested interests is critical.

Take water as an example: the goal articulated in the Millennium Summit is to cut by half by 2015 the number of people that lack access to water for drinking and sanitation. Attaining this objective translates into providing potable water for an additional 1.5 billion people and 2.1 billion people in the case of sanitation, the annual cost of which, according to *The Economist* (15 May 2005: 75), is estimated to be $1.7 billion and $9.3 billion respectively at a bare minimum, but going as high as $40 billion for an acceptable standard. This, the author points out, would be an expenditure that would yield

> an impressive rate of return by any measure, the benefits being social in the sense of avoiding illnesses attributable to dirty water, bad or non-existent sanitation and poor standards of hygiene stemming from lack of access to clean water.

This scale of increase, imprecise as it may be, is not attainable unless and until there is a large increase in official development assistance (ODA) and in private investment flows. A minimum estimate of the capital needed adds up to an amount that would call for a four-fold to five-fold increase in the total of foreign aid that now flows at an annual rate of about $50 billion, or, to take another measuring rod, adds up to roughly a doubling of the total flow of official aid, private bank lending and investment. While private capital flows to the developing countries have been increasing over the past few years, the overwhelming proportion of this capital flow is being directed to a very few developing countries under the aegis of multinationals that are investing for ventures with short payback periods. The fact is that too little of this capital is spent on the type of research, infrastructure and services that respond to the priority needs of the developing countries. Since it is not in the short-term financial interests of donor governments and private investors to be concerned with this, the systemic rules of the game provide insufficient incentives to do much about it.

But, lest it be thought that the financial aspect is the only way that vested interest thinking manifests itself, it needs to be stressed that there is also resistance to change by those who have a vested interest in maintaining positions of political/bureaucratic power whether in the public sector or the private. This resistance is a factor that goes part way to explain why there is support by some and resistance by others in the United Nations System. That is called upon to play the role of the host organization for the series of megaconference. There are many niches for the exercise of power in a large complex organizational system with a multifaceted mandate such as the United Nations System that is comprised of many separate (almost completely) autonomous agencies in which the head of each operates like a feudal lord, always on the alert to protect or to expand the agency's turf. (For example, over a dozen international organizations are concerned explicitly with issues related to water management so this jurisdictional issue is one demanding alertness and toughness.) The United Nations' Secretary-General is not vested with the requisite power to forge collaborative initiatives between these United Nations agencies. Persuasion is a weak force, but it is all that seems available under the "constitutional" arrangements that established each of the various agencies that comprise the United Nations *as a system.* The specialized agencies are constitutionally distinct entities and, as one author (Taylor 2003: 19) noted, each one of them has "a strong urge to go their separate ways and have the power to do so…It is a stubbornly polycentric system".

The member countries of the United Nations System have long recognized the need for coordination and, indeed, in the late 1960s, the UNDP's Inter-Agency Consultative Board commissioned Sir Robert Jackson to study the operation of the United Nations with regard to its mandate to promote development. His report was very critical in noting the United Nations' "non-political weakness" in not having the capability to work as "a highly coordinated unit under governmental control via the General Assembly" (Jackson 1969). He went so far as to liken the United Nations to

> some prehistoric monster, incapable of intelligently controlling itself, not because it lacks intelligent and capable officials, but because it is so organized that managerial direction is impossible.

Two decades later, the United Nations' Secretary-General attempted once again to overcome the lack of coherence that characterizes the structure of the United Nations as a system of agencies (see Box 7). The UNDP Administrator was appointed to act as a coordinator. This attempt to address the problem of coordination could not succeed to any appreciable degree and for very good reasons. Merely to sketch its operational dynamic reveals the difficulty of overcoming the vested interest to achieve effective coordination.

Not much, or not enough, seems to have changed since these authors and others have written and spoken about the issue. Over the decades there obviously has been formidable resistance to change. Yet without radical institutional changes that would reduce complexity, the role of this variant of vested interests will continue to frustrate efforts to "do the right thing".

But the direction of institutional change is towards increasing complexity. The Millennium Development Conference provides an illustration of this. The scope of the commitments was enlarged and was also made complicated by resolutions of the prior conferences. It should suffice to cite this as an example. A central issue about which there is little disagreement is, namely, the concept of human rights. Almost all the United Nations' specialized agencies have laid claim to responsibility for one or more aspects, and were helped to do so by the resolutions emanating from prior well-publicized megaconferences. The net result has been overlapping in terms of themes and of proposals (see Box 8).

As Michael Schecter observed in his article on the difficulty of achieving effective follow-up of these conferences (Schecter 2001: 218–222),

> it is no easy task to maintain a clear focus on the specific issues at the core of each conference, while at the same time, ensuring that all the conferences advance a comprehensive view of development. The agenda of the conference have tended to take a cross-sectoral approach (but) the United Nations System had tended to be organized along sectoral lines...As the United Nations' organs, bodies and programmes proliferated, so has the need for greater coherence within the system…The sweeping, cross-sectoral approach adopted by the global conferences has made the need for coherence even greater.

Box 7

The Coordination Conundrum

The founders of the United Nations Charter, trying to improve on the mechanism of the League of Nations for overseeing the economic and social institution, agreed to establish the Economic and Social Council (ECOSOC) to carry out this specialized function. But they did not give ECOSOC the necessary powers to manage effectively (as) it was only empowered under Article 61–6 of the Charter to issue recommendations to the specialized agencies, that are not only self-contained constitutionally, but are self-sustaining financially and not subject to direct United Nations control. Thus, as Sir Robert Jackson noted:

> Historically there has been no organization capable of defining a coherent overall agenda or coordinating and managing the wide range of economic and social activities which were carried out beneath the U.N. umbrella. The Administrative Committee for Coordination (ACC) which was intended to function as the main coordinating mechanism has generally failed as its members, the Agency heads, used it to defend their territories rather than agree (about) its management.

Another author, Martin Hill, in his book on coordinating the economic and social activities of the United Nations (Hill 1978: 95) was very categorical about this issue as it presented itself, a decade after Sir Robert Jackson had sung the same plaintive song:

> There exists no means of harmonizing the thinking of the executive heads and the senior staff of organs concerned with central policy issues, such as UNCTAD, UNIDO, UNDP, and directing it towards problems facing the international community and towards possible initiatives that the United Nations might usefully take.

There have been many attempts to address the problem posed by this lack of coherence, two of which involving institutional and procedural changes are illustrative. The process to achieve coordination has involved several steps:

- The establishment of functional commissions such as the Commission on the Status of Women, the Commission on Sustainable Development, the Commission on Human Rights and the Commission on Population and Development.
- The reinvigoration of the Economic and Social Council (ECOSOC) that operates under the authority of the General Assembly in trying to harmonize the agendas and work programmes of the functional commissions.
- Establishing a coordinated follow-up with regard to the implementation of the goals of these conferences, especially on the twelve themes common to these conferences, one of which was the provision of "basic social services for all: primary health care, nutrition, education, safe water and sanitation, population and shelter" (United Nations, E1995/86, 25–38, Table 1).

Box 8

Human Rights as a case in point re Conference Theme Overlap

The World Conference on Human Rights in 1992 defined the scope as inclusive of economic, social and cultural rights, the rights of women and children, and the right to development. This laid the foundation for the ambitious Agenda 21 that was formulated in 1992 at the United Nations Conference on Environment and Development that went beyond specifying action on environmental issues to include proposed action with regard to poverty, children, women, education, private sector involvement, etc.

The United Nations International Conference on Population and Development in 1994 was equally ambitious in expanding its scope in relating population to economic development and the environmental goals of the 1992 conference. The 1995 World Summit for Social Development expanded the scope yet again by including social integration as well as poverty and employment. The United Nations Conference on Women in that same year identified 12 critical areas of concern from poverty to armed conflict in addition to health, the environment and human rights. And so it went from year to year from conference to conference to include food and housing and other aspects that became the foci of other conferences.

The coordination of proposals and follow-up on each of these themes has proven to be a difficult task as has become evident in the reports for the special sessions that are 5-year post-conference reviews of progress (United Nations, E1998/19 and E1999/11). The further complicating factors in the institutional process, namely the need to coordinate the involvement of the various specialized agencies of the United Nations System, has been addressed by establishing a special entity for that purpose, the Administrative Committee on Coordination (ACC). In turn the ACC in 1995 established three inter-agency task forces (IATFs) whose roles are to coordinate three broad themes arising from the conferences:

- environment for social and economic development;
- employment; and
- basic social services for all (that includes the provision of water).

There is, as yet, little evidence that this complicated process is working or could work effectively to achieve the desired results on a significant scale. Overlaid on all this, as if to acknowledge that the coordination arrangements cannot be expected to work well, an entity called the Millennium Project has been established as "an independent advisory body". Beyond the issuance of impressive reports, the contribution of this institutional add-on remains to be seen. But, in any case, some basic questions remain: was a megaconference needed to have these reports commissioned and have they advanced the day of follow action?

The sad fact is that the act of signing protocols has too often been a ritual to be followed up by inaction or by half-hearted scarcely effectual steps (see Box 9). Writing on the issue of follow-up, Richard Jolly, a person steeped in United Nations lore through his long involvement with several United Nations agencies, has suggested that the prospects for effective follow-up might be more hopeful if there were the following factors at play (Jolly 1997):

> strong determined national and international leadership...(along with) political and social mobilization...(to undertake) doable low-cost strategies...(to achieve) a focused set of priorities.

Where are these attributes likely to be found? The very limited success of these conferences to achieve an action programme with the requisite financing can only in small part be attributable to the absence of *strong and determined national and international leadership*. There are the forces of inertia, risk-aversion and vested interests to contend with. Under these circumstances, it is important to recognize that the modality of United Nations conferencing on a global scale would not likely be effective as a means of tackling the host of issues that desperately need to be addressed. But political vested interests will want to play the charade of staging well-publicized megaconferences full of sound and fury but signifying little else.

So what is to be done?

Box 9

The follow-up to the Convention to Protect Wetlands provides an example when, after a decade of failure to advance towards the declared goals, the United Nations nonetheless pops off the cork to celebrate its "achievements" with a United Nations day. The same story could be told with regard to the dismal record of the 1994 Convention to Combat Desertification of 1994 that *celebrated* its tenth birthday with another special United Nations day though the pace of desertification has accelerated, prompting a spokesman for the United Nations department overseeing the programme to characterize the situation as "a creeping catastrophe" as each year since the Convention was signed about 3500 square kilometres have turned to desert and it is estimated that by 2025 two-thirds of arable land in Africa, one-third in Asia and one-fifth in South America will have disappeared leaving about 135 million people at risk or accelerating their exodus from rural areas to urban centres that are already strained to provide adequate clean water and sanitation facilities.

2.5 If No Mega Conferences, What Are the Alternatives?

I'm getting married in the mornin'
Ding dong! the bells are going to chime .
Kick up a rumpus, but don't lose the compass
And get me to the church on time.
For Gawd's sake, get me to the church on time!
—lyrics from *My Fair Lady*

Both the theory and practice of United Nations conferencing on a mega scale would seem to indicate that this mode of operating to address global problems or crises could not be relied upon to get us to the figurative "church". And, given its cumbersome structure and procedures, such conferences could not be, and should not be, counted on to overcome the obstacles to change and, in any case, to do so on time. Thus, there emerges a wide gap between what is promised and what is delivered, between rhetoric and reality.

Where does this leave us?

When a serious problem and/or an impending crisis is of a nature and scale that precludes any single nation from having any hope of success in tackling it alone, there is an understandable inclination to convene a meeting of interested parties. But several questions arise: need these meetings be global and, therefore, of mega size and scope? Need the outcome of discussions be couched in terminology that is cliché-ridden, that is to say, operationally vacuous?

An issue related to water management can provide an illustrative case: There are calls for the establishment of a World Water Institute (Kirpich 2004) to lead the research to more effectively address such phenomena as the excess of water (flooding), the dearth of water (desertification), the contamination of water, the conflicts in the allocation of water to meet the urban explosion in the demand for water for households, for electric power production and for agriculture use. All these are of great concern in many parts of the world and in that sense the problems have global dimensions. The challenge is complicated by the crisis aspect that adds the element of urgency:

- How fast could scientific, technological and policy research be geared up to significantly reduce the unacceptable death toll of about 12 million people each year that is directly attributable to waterborne diseases stemming from a lack of access to uncontaminated water?
- Could the amount of water available for agricultural purposes be increased sufficiently over the next quarter century to meet the anticipated increase of 80% in the need for food?

But, on the contrary, it is relevant to ask:

- Could/should these problems that are couched in global terms be more effectively tackled at the level of national or regional governance (river basin or metropolis) since its resolution would seem to be well within the purview and the capacity of a national or sub-national government?

However, if the answer is that a research programme is best underwritten and undertaken by an international body for financial and other reasons such as the size and frequency of risk of disaster and urgency of response, a case could be made that the appropriate measures should be within the purview of global governance. The research in water management could be very helpful if focused on technology for finding water sources underground, for desalination plants, for pumping, for transporting water and for determining the best ways in which water should be used in households, in industry and on the farm. For example, research has already brought down the cost of desalination to less than 50¢ per m^3, or less than $1.80 per 1,000 gallons and the minimum scale of the units has been radically reduced so that the cost effectiveness issue is being resolved. The outreach to spread the knowledge of the research findings and accelerate their application through the dissemination of literature and establishment of training programmes would also call for institutional arrangements and programmes that are international in their financing and in their implementation as "best practices".

The question that follows is: If a mega scale conference is not an effective way to launch such research initiative, what is the alternative?

Three initiatives come to mind that were conceived and implemented without the need for a megaconference to launch them, one of which is rather recent and two, rather dated but, nonetheless, suggestive and pertinent for today:

1. The first is the agreement called the Montreal Protocol that was launched by a process involving small-scale meetings and a conference that were all sharply focused on the issue of ozone depletion as an environmental challenge of global dimensions. Here the objective of the meetings was to devise an agreement among all nations on the why and how and when of reducing the emission of chlorofluro-carbons (CFCs) into the atmosphere. The reasons why the Montreal Protocol has been a success story that is still in progress should be identified to reveal under what conditions Union Nations conferences could hope to succeed. The factors were the following:

- It was more a series of relatively small meeting in terms of participants and publicity;
- Its recommendations to reduce the amount of CFCs released into the atmosphere was based on accepted scientific knowledge;
- There was a feasible alternative to CFCs in terms of cost and ease of application by industries in both the industrialized and developing countries;
- The results could be monitored and nations could be encouraged and pressured to undertake the necessary regulatory regime by the publication of periodic reporting on shortfalls, thereby exerting moral suasion. (Pending a form of "world government" able to enforce rules of "good behaviour". Reliance has to be placed on a process of shaming non-compliant nations.)

2. The establishment of the Consultative Group on International Agricultural Research (CGIAR). The idea was initiated by three non-governmental foundations (Kellog, Rockefeller, Ford). The World Bank and the United Nations' Food and Agricultural Organization (FAO), in 1970, enabled the transformation of that idea

into an operational entity for undertaking research to increase food output and improve policy-making on all issues related to agricultural production and consumption (see Box 10).

Box 10

CGIAR is an alliance of governments, regional and international organizations, donor institutions and research centres that mobilize funds and expertise for both technical and policy advising for the benefit of poor farmers in developing countries. It now supports 18 research centres around the world, all but one in the developing countries. (The policy entity is the International Food Policy Research Institute (IFPRI) that is located in Washington, DC.)

For a fuller exposition, in addition to CGIAR'S annual reports, see Anderson et al. (1988) and Miller (1992).

It is estimated that the increase in food production attributable to CGIAR as a way of organizing and financing research more effectively has yearly enabled the feeding of more than 1 billion people.

The report, *CGIAR Annual Meetings 2004: In Search of Solutions for the Farmer of the XXI Century*, reveals that each dollar that has been invested through its research system has generated nine additional dollars in increased food production, and that income *per capita* in the poorest economies of the world would not have increased by about 7% had there not been the active presence of the CGIAR. Its dynamism is typified by the recent launching of a $120 million dollar research consortium, known as the Challenge Programme on Water and Food, to investigate how more food can be produced with less water, that is explore new technologies to optimize the use of water in agriculture that presently accounts for 70 to 90% of water use.

It seems relevant to note in this connection that neither the concept nor the establishment of CGIAR arose from a conference at all. Sometime in 1969 the heads of the Ford, Kellogg and Rockefeller foundations made a decision to arrange a meeting with the President of the World Bank, Robert McNamara, to discuss how they could meet the demand for funds to support agricultural research that was overwhelming them. It was a time of crisis in the sense that during the 1950s and 1960s there had been famines in India and China that took a toll of hundreds of millions of lives and the need for increased agricultural output was obvious. They decided on the establishment of CGIAR as a means of using whatever funds were available in a more effective manner. The result is, between 1970 and today, while global population doubled, food production tripled and over time both India and China became net exporters of food.

3. The establishment of the Manhattan Project to build the atom bomb was a response to a crisis or, more accurately, a response to head off a potential disaster that was deemed to be possible or even likely. The threat was the possibility of the

Nazi regime succeeding in producing the atom bomb with all its dire implications. Though in a much different context with the global warming hypothesis as the impending global threat, voices are being heard that allude to the Manhattan Project as a relevant precedent for preventive action (see Box 11).

Box 11

An eminent professor of physics at New York University, Dr Martin Hoffert, has stated:

> the country needs to embark on an energy research programme on the scale of the Manhattan Project ...or the programme that put a man on the moon. Maybe six or seven of them operating simultaneously.....We should be prepared to invest several hundred billion dollars in the next 10 to 15 years.

Dr Arthur Nozik, a senior research fellow at the National Renewable Energy Laboratory, has echoed Professor Hoffert. In an article in *The New York Times* of 4 November 2003, he is reported to have stated that:

> We need something like a Manhattan Project or an Apollo programme to put a lot more resources into solving the problem. It is going to require a revolution, not an evolution.

And now columnists are adding their voice: recently *The New York Times* columnist and author, Thomas L. Friedman, posed the question: "Why didn't the (United States) administration ever use 9/11 as a spur to launch a Manhattan Project for energy independence and conservation?"

The context for the launching of the Manhattan Project that was rationalized on the basis of the precautionary principle is somewhat akin to the current global anxiety about climate change due to the warming effects of ever higher carbon emission levels. The reports of the Inter-governmental Panel on Climate Change (IPCC 1994), which was established to follow-up a resolution of a United Nations megaconference, are the bible with its warning of a possible and likely dire denouement for the humanity if the precautionary principle is not taken seriously enough.

What is being suggested by these three examples of relatively successful initiatives is that the hope of being effective in making significant changes is to advance energetically on a few *salient* issues that are feasible in the sense of the likelihood of being endorsed and being implemented within the prevailing global economic/financial system. The outcome of the smaller sharply focused type conferences is more likely to lead to actionable initiatives as contrasted with the approach of the traditional elephantine exercises of mega conferencing that hardly go further than securing pledges to toothless protocols or agreements. We should take heed of the 103 Nobel Laureates who stated in their Millennium Manifesto of December 2000:

> to survive in the world we have transformed, we must learn to think in a new way.

Bibliography

Alger C (ed) (2003) The Future of the United Nations System: Potential for the 21st Century. United Nations University Press, Tokyo

Anderson JR, Herdt RW, Scobie GM (1988) Science & Food: The CGIAR and Its Partners. World Bank, Washington DC

Ayton-Shenker D (ed) (2002) A Global Agenda: Issues Before the 57th General Assembly of the United Nations. Chapter 4: Global Resources. An annual publication of the United Nations Association of the USA. Rowman & Littlefield, Lantham Md

Barnett M, Finnemore M (2001) The Politics, Power and Pathologies of International Organizations. In: Martin LL, Simmons BA (ed) International Institutions: An International Organization Reader. MIT Press, Cambridge, MA, pp. 699–732

Barrett B. van Ginkel H, Court J, Velasquez J (2004) Human Development and the Environment: Challenges for the United Nations in the new Millennium. UNU Millennium Series, United Nations University Press, Tokyo

Barrett S (1999) Montreal vs. Kyoto: International Cooperation & the Global Environment. In: Kaul I, Grunberg I, Stern MA (eds) Global Public Goods: International Cooperation in the 21st Century. Oxford University Press, New York & London, pp. 192–219

Biswas AK, Uitto J (2004) Water for Urban Areas: Challenges and Perspectives. UNU Series on Water Resources Management & Policy, UNU Press, Tokyo

Biswas AK (2004) Integrated Water Resources Management: A Reassessment. Water international 29(2): 248–256

Biswas AK (2004) From Mar del Plata to Kyoto: an analysis of global water policy dialogue. Elsevier 14: 81–88

Boutros Boutros-Ghali (1999) Unvanquished: AUS-United Nations Saga. Random House, NY

Bruyn S (2001) Civil Associations Towards a Global Civil Economy. In: Harris J, Wise T, Gallagher K, Goodwin N (eds) A Survey of Sustainable Development: Social and Economic Dimensions. Island Press, Washington DC, pp. 289–292

CGIAR (2004) Annual Meetings 2004: In Search of Solutions for the Farmer of the XXI Century. Washington, DC

Chasek P (ed) (2004) The Global Environment in the 21st Century: Prospects for International Cooperation. United Nations University Press, Tokyo

Cooper A (2004) Tests of Global Governance: Canadian Diplomacy and the United Nations World Conferences. United Nations University Press, Tokyo

Copenhagen Consensus Project (2004) Global Crises, Global Solutions. Cambridge University Press

Cox R, Jacobson H (1973) "Framework for Inquiry" and "The Anatomy of Influence." In: Cox R, Jacobson H (eds) The Anatomy of Influence: Decision-making in International Organization. Yale University Press, New Haven, CT pp. 1–36, 437–465

Doyle T (1998) Sustainable Development and Agenda 21: The Secular Bible of Global Free Markets and Pluralist Democracy. Third World Quarterly 19(4): 771–786

Elliott L (2002) Global Environmental Governance. In: Wilkinson R, Hughes S (eds) Global Governance: Critical Perspectives. Routledge, London & New York, pp. 57–74

Essex C, McKittrick R (2002) Taken by Storm: the Troubled Science, Policy and Politics of Global Warming. Key Porter Books, Toronto

Fredriksson PG (1999) Trade, Global Policy and the Environment. World Bank Discussion Paper No. 402, Washington, DC

Freeman MA (2002) Environmental Policy Since Earth Day 1: What Have We Learnt? Journal of Economic Perspectives 16(1): 125–146

French H (1995) Partnership for the Planet: An Environmental Agenda for the United Nations. Worldwatch Institute, Washington, DC

French H (2000) Coping with Ecological Globalization. State of the World 2000. Worldwatch Institute, Washington, DC

Heal G (1999) New Strategies for the Provision of Global Public Goods: Learning from International Environmental Challenges. In: Kaul I, Grunberg I, Stern M (eds) Global Public Goods: International Cooperation in the 21st Century. Oxford University Press, New York, pp. 220–239

Henderson H (1996) Creating Alternative Futures: the End of Economics. Putnam, New York

Hill M (1978) The United Nations System: Coordinating its Economic and Social Work. Cambridge University Press, Cambridge, UK

IPCC (Intergovernmental Panel on Climate Change) (1994) Climate Change 1994: Radiative Forcing of Climate Change. Cambridge University Press, Cambridge (and subsequent reports issued by the IPCC)

International Institute for Applied System Analysis (IIASA) (1996) Water Resources: Good to the last drop? Options, Summer

Jackson R (1969) A Study of the Capacity of the United Nations System. United Nations, Geneva

Johnson B, Duchin F (2000) The Case for the Global Common. In: Harris J (ed) Rethinking Sustainability: Power, Knowledge and Institutions. University of Michigan Press, Ann Arbor

Jolly R (1997) Human Development: The World after Copenhagen. 1996 JW Holmes Memorial Lecture, ACUNS Reports and Papers, No. 2 (Providence, Rhode Island's Academic Council on the United Nations System)

Jolly R (2001) Implementing Global Goals for Children: Lessons from UNICEF Experience. In: MG Schecter (ed) United Nations-sponsored World Conference: Focus on Impact and Follow-up. United Nations University Press, Tokyo, pp. 10–28

Jordan A, Voisey H (1998) Institutions for Global Environmental Change. Global Environmental Change 8(1): 93–97

Kaya Y, Yokobori K (eds) (2004) Environment, Energy and the Economy: Strategies for Sustainability, United Nations University Press, Tokyo

Kirpich P (2004) Water Management: The Key Role of the International Agencies. Water International 29(2): 243–247

Kulshreshtha SN (1994) World Water Resources and Regional Vulnerability: Impact of Future Changes. IIASA Report. Luxemburg, Austria

Lichenstein C (1986) The United Nations: Its Problems and What to Do About Them. Heritage Foundation, Washington, DC

Lipietz A (2001) Enclosing the Global Commons: Global Environmental Negotiations in a North-South Conflictual Approach. In: Harris J, Wise T, Gallagher K, Goodwin N (eds) A Survey of Sustainable Development: Social and Economic Dimensions. Island Press, Washington, DC, pp. 107–114

Manes C (1990) Green Rage: Radical Environmentalism and the Unmaking of Civilization. Little Brown, Boston

Martin L, Simmons B (2001) Theories and Empirical Studies of International Institutions. In: Martin LL, Simmons BA (eds) International Institutions: An International Organization Reader. MIT Press, Cambridge, MA, pp. 437–465

Martin L (2001) Interest, Power and Multilateralism. In: Martin LL, Simmons BA (eds) International Institutions: An International Organization Reader. MIT Press, Cambridge, MA, pp. 37–64

Matthews WH (1980) Moving Beyond the Environmental Rhetoric. Mazingira: International Journal for Environment and Development 4(2): 6–15

McInerney-Lankford S, Salman MAS (2004) The Human Rights to Water: Legal and Policy Dimensions. Oxford University Press, NY

Meinzen-Dick RS, Rosegrant MW (2001) Overcoming Water Scarcity and Quality Constraints. IFPRI, Focus No. 9, Brief No. 1

Miller M (ed) (1961) Resources for Tomorrow Conference, Papers and Proceedings (3 volumes). Queen's Printer, Ottawa

Miller M (1983) The Challenge of the Energy Transition: the United Nations Response. In: El-Hinnawi E, Biswas MR, Biswas AK (eds) New & Renewable Sources of Energy. Tycooly International Publishing, Dublin, pp. 1–8

Miller M (1992) Getting Grounded at Rio. Ecodecision: Environmental Policy Magazine 5: 55–57

Miller M (1993) Sustainability and the Energy/Environment Connection: Overcoming Institutional Obstacles to 'Doing the Right Thing'. In: Siddayao C (ed) Investing in Energy and the Environment, World Bank, Washington, DC, Chapter 4, pp. 159–204

Miller M (1994a) The Role of Large-Scale United Nations Conferencing: Promoting Bio-energy & Environment Programs (Working Paper No. 94–57). University of Ottawa, Faculty of Administration, Ottawa (ISSN 0701-3086)

Miller M (1994b) Reconciling Developmental and Environmental Objectives: Necessary Conditions and Strategic Options, University of Ottawa, Faculty of Administration, Ottawa (ISSN 0701-3086). Working Paper No. 94–55

Miller M (1995) The Environment Policy Challenge. In Sharma S (ed) Development Policy. St. Martin's Press, New York, Chapter 8, pp. 128–155

Miller M (1997) High-Tech to the Rescue? The Role H-T could Play involving the Rural Poor in the Knowledge Economy. Working Paper No. 97–25. School of Management, University of Ottawa (ISSN 0701-3086)

Miller M. (2005) Sustainable Development: A Flawed Concept. In: Biswas AK, Tortajada C (eds) Appraising Sustainable Development: Water Management and Environmental Challenges, Oxford University Press, pp. 18–64

Mitchell A (2004) Dancing at the Dead Sea: Tracking the World's Environmental Hotspots. Key Porter Books, Toronto

Munasinghe M (1993) Issues and Options in Implementing the Montreal Protocol in Developing Countries. In: Munasinghe M (ed) Environmental Economics and Natural Resources Management in Developing Countries. Committee of International Development Institutions on the Environment (CIDIE), World Bank, Washington, DC

Nordhaus WD (1990) Greenhouse Economics: Count before You Leap. The Economist (July 7) pp. 21–24

Olsen M (1965) The Logic of Collective Action. Harvard University Press, Cambridge, MA
Ono M (2001) From Consensus-building to Implementation: The Follow-up to the United Nations Global Conferences of the 1990s. United Nations-sponsored World Conferences: Focus on Impact and Follow-up, ed. Schecter M.G, U.N.U. Press, Tokyo, pp. 169–178
Organization for Economic Development (OECD) (1979) Interfutures: Facing the Future, Mastering the Probable and Managing the Unpredictable. OECD, Paris
Padbury P (1993) UNCED and the Globalization of Civil Society: Lessons for United Nations Reform. Mimeo, Ottawa
Pearce D (1996) Towards a Sustainable World. Scientific American (Special Issue: Managing the Earth) 3(4)
Pearson CS (1985) Down to Business: Multinational Corporations, Environment and Development. World Resources Institute Study No. 2, Washington, DC
Reddy A, Williams R, Johansson T (1997) Energy after Rio: Prospects and Challenge. UNDP, New York
Repetto R, Austin D (2001) The Costs of Climate Protection: A Guide for the Perplexed. In: Harris J, Wise T, Gallagher KP, Goodwin N (eds) A Survey of Sustainable Development: Social and Economic Dimensions. Island Press, Washington, DC, pp. 216–219
Riggie JG (1972) Collective Goods & international Collaboration. American Political Science Review No. 66
Righter R. (1995) Utopia Lost: the United Nations and World Order. 20th Century Fund Press NY
Rijsberman F (2004) The Water Challenge. International Water Management Institute, Columbo, Sri Lanka (commissioned by the Copenhagen Consensus Project)
Rittberger V (ed) (2004) Global Governance and the United Nations System. United Nations University Press, Tokyo
Rosenfeld SS (1994) Cairo Mandate (Conference). Washington Post, September 16th
Rosegrant MW (1997) Water Resources in the 21st Century: Challenges and Implications for Action. IFPRI Discussion Paper No. 20
Rotstein A, Harries-Jones P, Timmerman P (1999) A Signal Failure: Ecology and Economy After the Earth Summit, Future Multilateralism: The Political and Social Framework, (ed) Schecter, M., UNU Press, Tokyo, pp. 101–135
Ruffing K (2002) Johannesburg Summit: Success or Failure? OECD Observer, No. 234 (October)
Sachs I (1997) Rio, Five Years Later: Against a Wintry Sky, a Few Swallows. Ecodecision 10
Sachs I (1980) Strategies de l'Ecodeveloppement. Les Editions Ouvrières, Paris
Sachs JD (2005) The end of Poverty: Economic Possibilities for our Times. The Penguin Press, New York, pp. 82–89, 210–225, 285–287
Schecter M (ed) (1996) Future Multilateralism: the Political and Social Framework. Macmillan, London, 1998
Schecter M (ed) (2001) United Nations-sponsored World Conferences: Focus on Impact and Follow-up. United Nations University Press, Tokyo
Shiva V (2001) Conflicts of Global Ecology: Environmental Activism in Periods of Global Reach. In: Harris J (ed) A Survey of Sustainable Development: Social and Economic Dimensions. Island Press, Washington, DC, pp. 365–367
Somavia J (1995) Post-Copenhagen: Personal Reflections. In: Advancing the Social Agenda: Report of the UNRISD International Conference. United Nations, Geneva

Strong M (2001) More is Not Enough. Scientific American (Special Issue: the Growth-Environment Dilemma) 3(4)

Taylor P (2003) The Social and Economic Agenda of International Organization in the Age of Globalization. International Organizations in the Age of Globalization, Continuum. New York & London, pp. 135–166

The Economist (2004) Understanding the global water crisis. London, May 13th

Tolba MK (1988) Evolving Environment al Perceptions: From Stockholm to Nairobi. United Nations Environment Programme (UNEP), Buttersworth, London

Tucker W (1980) Environmentalism: the Newest Toryism. Policy Review 14: 141–152

United Nations Research Institute for Social Development (1997) Public Meeting Two Years After Copenhagen. UNRISD/Conf/93/3, Geneva (December)

UNESCO (2003) The World Water Report. Paris

United Nations (1995) The Report of the Secretary-General on coordinated follow-up to major international conferences in the economic, social and related fields. (E/1995/86)

United Nations (1995) Report of the Secretary-General on coordinated follow-up to major international conferences in the economic, social and related fields. (E/1995/86),

United Nations (1996) ECOSOC, Consultative Relationship between the United Nations and Non-governmental Organizations. (1996/31, Part I)

United Nations (1998) Report of the Secretary-General on integrated and coordinated implementation and follow-up of major United Nations conferences and summits. (E/1998/19)

United Nations (1999) Report of the Secretary-General on Meeting on Basic Indicators: Integrated and coordinated implementation and follow-up of major United Nations conferences and summits. (E/1999/11)

United Nations (2005) Investing in Development: A Practical Plan to Achieve the Millennium Development Goals and the reports of the 10 Task Forces of the Millennium Project

Von Braum J, Swaminathan MS, Rosegrant MW (2004) Agriculture, Food Security, Nutrition and the Millennium Development Goals. IFPRI Annual Report Essay 2003–2004. IFPRI, Washington, DC

Wente M (2002) The Kyoto-speak Brainwashes. The Globe & Mail, Toronto (December)

Wijkman P (1982) Managing the Global Commons. International Organization No. 36. Washington, DC

Wilkinson R (2002) Global Governance: a preliminary interrogation. In: Wilkinson R, Hughes S (eds) Global Governance: Critical Perspectives. Routledge, London, pp. 1–14

World Bank (1999) A Strategic View of Urban and Local Government Issues: Implications for the Bank. Washington, DC, p. 6

World Bank (2002) The Environment and the Millennium Goals. Washington, DC

World Bank (2004a) Overview: From Vision to Action. The Global Monitoring Report 2004. Washington, DC

World Bank (2004b) Rising to the Challenge: the MDG for Health. Washington, DC

Young O (1989) International Cooperation: Building Regimes for Natural Resources and the Environment. Cornell University Press, Ithaca, NY

Young O (2001) Political Leadership and Regime Formation: On the Development of Institutions in International Society. In: Martin LL, Simmons BA (eds) International Institutions: An International Organization Reader. MIT Press, Cambridge, MA, pp. 9–36

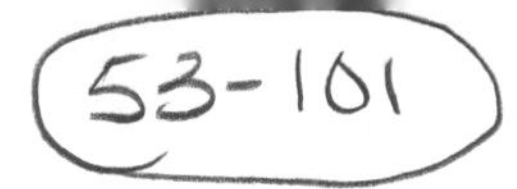

3 Global Water Initiatives: What Do the Experts Think?

Robert G. Varady and Matthew Iles-Shih

3.1 Introduction

Global water initiatives (GWIs) are institutions whose fundamental purpose is to advance the knowledge base regarding the world's inland water and its management. Additionally, since the 1980s, the core aim of many GWIs has expanded to include an active social and policy component. Thus, the mandate of many of these initiatives now includes attempts to improve access to potable water and sanitation across the globe.

In view of the great diversity of water-related issues, institutions of several types have helped to generate knowledge and create social change. Four such types of water initiatives are addressed in this chapter, including professional societies, designated time periods, organized events, and inter-governmental and non-governmental organizations.

Although global water initiatives have existed for more than a century and their numbers have increased palpably since World War II, surprisingly little has been written on their collective activities and impact. To attempt to redress the paucity of research on global water initiatives, Varady began a study of this phenomenon in 2003, a two-part inquiry. On one front, he collected primary and secondary written sources on the initiatives' origins, objectives, leaders and workings. From this information and from conversations with knowledgeable individuals, a contextual framework was constructed. This framework for the evolution and significance of global water initiatives was presented at several conferences, seminars and workshops (see, e.g., Varady 2003, 2004). The main features of this exploration appear in Section 3.3.

Concurrently, to help answer key questions on the genesis, operation and influence of the most significant initiatives, and to better understand the nature of their interactions, Varady surveyed about 120 influential participants and knowledgeable individuals, including officials at nearly 40 international water-related institutions. The inquiry seeks to determine the degree and effects of institutional overlap; identify the most significant actions taken by and overall impact of GWIs and draw explicit lessons learned by participants in the course of their work with GWIs.

An intuitive working hypothesis, based on conversations, readings and early impressions, is that the numerous existing global water initiatives have frequently duplicative aims and have over-proliferated. Under such a hypothesis, one would expect that experts in the field – a sophisticated and generally sceptical set of informants – would tend to minimize the salutary influences of GWIs and perhaps advocate their consolidation or selective elimination.

This is a report on the survey. Accordingly, in the sections that follow, the authors describe the two survey instruments, outline the methodology and tools employed, present selected findings drawn from an analysis of completed questionnaires[1] and offer some discussion and conclusions on the most salient and meaningful observations. In the process, the validity of the hypothesis is assessed.

3.2 What Are Global Water Initiatives and Why Study Them?

As the editor of the journal *Water Policy* has written, "the history of social organization around river basins and watersheds is humanity's richest record of our dialogue with nature" (Delli Priscoli 1998: 623).[2] But throughout human history, the instruments available to nation-states have remained largely inadequate to handle global institutional problems. Not surprisingly then, it is only over the past few decades that scientists, government officials and world leaders have come to realize that water is a key resource whose availability, quality and effective management are central to assuring human health, prosperity and peace.

The immediate post-World War II period was marked by large, capital-intensive development projects. Then, beginning in the mid-1960s, partly because of rapidly increasing population and partly due to growing fears of conflict over water, international attention began to turn to the core issue of water policy (Wolf 1998). The decades since the early 1950s have featured concerted, organized activity intended to improve the understanding of and enhance access to the world's water resources.

As Figure 3.1 shows, one of the affects of these efforts has been the emergence and proliferation of a montage of water-related associations, programmes and organizations – what the authors refer to as *global water initiatives*. However, because these institutions have sprung from numerous and often divergent sources, attempts to develop innovative and practical observations and recommendations have sometimes been frustrated by the sheer number of voices and diversity of approaches continually emanating from this dynamic institutional "ecosystem".

The complete mosaic of global water initiatives remains poorly understood. To what extent do these initiatives constitute a well-defined network with clearly articulated links, traceable influences and unified purpose? Or, as some have rightfully asked, are the various efforts independent, poorly connected, even competing

---

[1] The data presented and interpreted in the following pages are drawn from a strategic sampling of responses to a larger set of questions. The results, while particularly appropriate for this book's specific theme, are not necessarily representative of the full range of data and findings of the larger analysis that is in progress.

[2] See Rodda and Ubertini (2004) for additional discussions on the role of water in the history of human societies.

enterprises? Are there, as many observers have suggested, *too many* concurrent initiatives? Or, as the respected thinker Malin Falkenmark has suggested, are we witnessing a generally wholesome instance of "institutional biodiversity"?[3] Underlying these questions, we can ask, as Asit Biswas has, whether the ensemble of water initiatives has made a palpable difference on the ground. In other words, "Would the world of water have been much different if [these initiatives] did not exist?"[4]

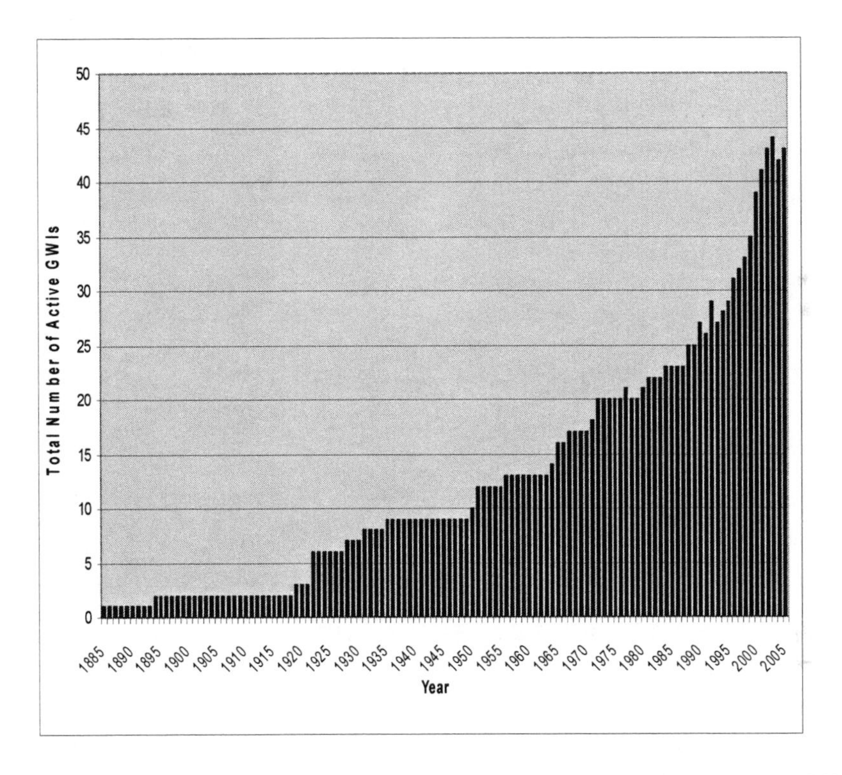

Fig. 3.1 Significant Global Water Initiatives: Population growth from 1885 to 2005 (Source: Authors' compilations)

[3] Interview of Malin Falkenmark by Robert Varady, Stockholm, Sweden, 3 May 2004.
[4] Personal correspondence with A. K. Biswas, 7 September 2002.

3.3 Background

3.3.1 The Roots of Water Consciousness and Its Internationalization

Professional Societies

Beginning in the late 1800s, the earliest modern organizations for those working in the field of water were professional societies. These groups were established to construct common intellectual spaces, share expertise, and stimulate and promote basic and applied research. By the mid-1950s, water scientists, engineers and managers had established respected, well-functioning and well-subscribed organizations, each pursuing the interests of its members and pulling in its own direction.

The oldest continuously operating professional water-related society *cum* interest group has been the Permanent International Association of Navigation Congresses (PIANC), which has existed for nearly 120 years. Perhaps the most comprehensive scientific organizations, dating to 1919 and 1931 respectively, are the International Union of Geodesy and Geophysics (IUGG) and the International Council for Science (first called the International Council of Scientific Unions, or ICSU). Both the IUGG and the ICSU have provided wide topical umbrellas that have accommodated hydrologists, hydrogeologists, hydraulic engineers and other water scientists and practitioners. As with other such transdisciplinary associations, specialists whose numbers were growing began to form their own groups. The earliest offshoots of the IUGG and the ICSU included such societies as the International Association of Hydrological Sciences (IAHS; formerly IASH) and the International Association of Theoretical and Applied Limnology (SIL), both formed in 1922, and the International Association for Hydraulic Research (IAHR), created in 1935. All these associations were born in the decades between the two world wars (George 2003). Table 3.1 lists some of the most significant water-related professional societies.

World War II and Its Aftermath: Multinationalism and Technology

The aftermath of World War II had a very significant influence in changing perceptions of water. In 1945 in the wake of the destructive 6-year upheaval, a strong sentiment for multinational approaches to avoiding new wars gave rise to the United Nations (UN). The signatories of the UN Charter recognized that many of the world's problems transcend political borders, and like issues of war and peace, are best addressed multilaterally (Victor and Skolnikoff 1999; Keohane et al. 1994; Udall and Varady 1993). Simultaneously, during the formative years of the UN, the western nations and the Soviet Union had at their disposal the potent technologies of the period. In the hubris of victory against Germany and Japan, the world powers brimmed with confidence over their ability to deploy the new technologies to transform society and adapt the landscape to human needs (Weiner 1992).

Table 3.1 Examples of influential international professional societies

Professional society	Date Established
Permanent International Association of Navigation Congresses (PIANC)	1885
Commission Internationale des Glaciers (Intl. Comm. on Snow and Ice)	1894
Intl Union of Geodesy and Geophysics (IUGG)	1919
Intl Association of Hydrological Sciences (IAHS; formerly IASH)	1922
Intl Geographical Union (IGU)	1922
Intl Assoc. of Theoretical and Applied Limnology (SIL)	1922
Intl Council for Science (ICSU; formerly Intl Council of Sci. Unions)	1931
IAHR (formerly Intl. Association for Hydraulic Research)	1935
Intl Union of Technical Associations and Organizations (UATI)	1951
World Irrigation and Drainage Congresses (triennial)	1951
Intl Association of Hydrogeologists (IAH)	1956
Committee on Water Research (COWAR)	1964
Intl Association for Water Law (AIDA)	1967
Intl Water Resources Association (IWRA)	1972
World Water Congress (triennial)	1973
Intl Hydropower Association (IHA)	1995
Intl Water Association (IWA)	1999
Intl Water Associations Liaison Committee* (IWALC)	2000
Intl Water History Association (IWHA)	2001

Source: Compiled by authors and partly based on personal communications with officials of some of the societies (e.g. IAHR, IGU, IHA and IWA).

* IWALC is unusual among these organizations in that it is a grouping of the chief executives of other water-related societies and has no general membership. While this admittedly stretches the definition of a "professional society", because IWALC's general aims are consonant with those of a society, the authors have chosen to include it in this category.

Nowhere was this new impulse to harness technology more clearly visible than in the realm of water. The years from 1945 to the late 1970s brought an unprecedented initiation of ambitious, large-scale waterworks such as dams, barrages, irrigation schemes and hydroelectric plants; river diversions and interbasin transfers; and wetlands-drainage and land-reclamation projects. Heralded as signals of 20th century progress, these enterprises underlined the centrality of water to society. During this time, numerous institutions arose to advocate one or another aspect of water.

At the same time, water scientists continued forming new professional societies whose aims reflected evolving priorities of water science and management. The International Union of Technical Associations and Organizations (UATI), the triennial World Irrigation and Drainage Congresses, and the International Association of Hydrogeologists (IAH), all originated in the early to mid-1950s. By this time, at the height of the Cold War, professional societies began to supplement their scientific and collegial goals with pursuit of certain social or political objectives, mostly in the nature of increasing dialogue and communication among colleagues. As an example, the International Association of Hydrogeologists, established in 1956, the year of the Soviet suppression of the Hungarian uprising, was strongly motivated to rectify the "virtual breakdown of relations between the countries of Eastern Europe and the West, together with the isolation of huge areas of Asia", which they saw as the cause of "enormous problems for international science and its practitioners" (Day 1999:1).

The International Hydrological Decade (IHD) and Its Origins

Post-war polarization not only isolated professionals from some of their counterparts, it created a gulf in the content of science. Ideological differences were reflected in the distinct schools of science and approaches to technology that began coalescing during this time. This prompted scientists, engineers, educators and UN officials to call for the designation of a unified and concerted global effort to gather and interpret data on the planet. The result was the International Geophysical Year (IGY), which lasted from July 1957 to December 1958 (Chapman 1959). IGY marked the first serious, sustained collaboration between Soviet and Western scientists and set the stage for other large-scale, focused and ideologically safe planetary science programmes.

The success of the IGY inspired other scientists, among them Michel Batisse of UNESCO, Raymond Nace of the United States Geological Survey, and Léon Tison of IAHS, three leading figures in the world of water. According to Batisse, the idea of a hydrological programme modelled on IGY first arose in mid-1960 in the United States, prompted by the National Science Foundation. Soon after, at an informal conversation at an IAHS symposium in Athens, Greece, Batisse, Nace, Tison and others began to explore the possibility of declaring an official designated time period for hydrology. By 1962, with a groundswell of support from other scientists, various quarters of the United Nations, and a number of key member nations, the formal process of planning for the International Hydrological Decade got under way (Batisse 2005:84). UNESCO, which did not at the time have a water-resources division, was nonetheless deemed best suited to arrange meetings, coordinate activities and provide multilateral leadership for planning and implementing the proposed Decade. Its major partner was a UN sister agency, the World Meteorological Organization (WMO). Following its November 1962 general meeting, UNESCO convened additional sessions aimed at broadening participation and re-examining and revising the original proposal to create an international decade. The recommendations of these meetings were adopted by the end of 1964 and UNESCO launched the International Hydrological Decade (IHD) at the start of 1965 (Korzoun 1991; Batisse 2003, 2005).

Most observers agree, and the surveys discussed below confirm, that the Decade, which ended in 1974, was a major boon to the field of water sciences as a whole and to understanding the hydrologic cycle in particular. At the outset the programme defined five main objectives: to collect hydrologic data, assess resources and budget balances, conduct research into problems, educate and train new personnel and facilitate information exchange. In the course of addressing these objectives, the Decade promoted scientific cooperation and substantially advanced the state of hydrologic knowledge. One of the by-products of the flurry of activities generated by the IHD was that it drew considerable attention to water issues.

One of the IHD's specific objectives, an inventory of the world's water balance, was accomplished not long after the end of the Decade with UNESCO's 1978 publication of *World Water Balance and Water Resources of the Earth* (Korzoun 1978). This comprehensive inventory provided previously unavailable basic data at different scales. More significantly, it offered the possibility of assessing the state of the planet's available water resources. In the process, the Decade prompted a succession of publications, such as an authoritative glossary and numerous monographs, papers, reports, educational materials and other documents. In addition, IHD convened at least 25 major international conferences, helped train technicians and generally raised the profile of the study of water and its problems.

The International Hydrological Programme (IHP)

The International Hydrological Decade's last official action was a scientific conference held in Paris in 1974. The final report of the IHD showed that more than 100 nations had taken part in the Decade, confirming the hypothesis that scientific cooperation would transcend political differences (Korzoun 1991). The immediate question raised by the IHD's apparent success was how to harness its momentum to carry forward its unfulfilled ambitions. Accordingly, the closing conference agreed to view the just-concluded IHD as the first part of an organic, long-term programme. UNESCO's 1974 General Conference took the lead in transforming the Decade into a periodically renewable institution called the International Hydrological Programme (IHP). Subsequent discussions centred on the eventual roles of two key UN agencies with sometimes overlapping water-related agendas, the World Meteorological Organization (WMO) and UNESCO. By agreement, IHP was housed at UNESCO, which agreed to provide the bulk of its budget. At that same time, UNESCO and WMO signed a cooperative accord that remains in force. In the 30 years since IHP's birth, WMO has been an important participant and partner in many of IHP's activities (Rodda 1991; UNESCO/WMO 1988).[5]

The goal of the new effort was similar to that of the Decade: *to strengthen the connections between scientific research, application and education in the realm of water*. Also like the IHD that preceded it, the International Hydrological Programme has been an engine of activity. It helped promote such influential conferences such as the 1977 UN Conference on Water in Mar del Plata, Argentina, as

[5] In particular, the WMO's broad-based and highly influential Hydrology and Water Resources Programme.

well as numerous scientific studies, training programmes and publications. But the IHP's most significant contribution may be its institutional centrality, persistence, and resilience. By offering a permanent forum for water-related interests, IHP has helped encourage multinational cooperation and stimulate innovative approaches to water science and management.

3.3.2 The Evolution of Global Water Initiatives

Following the previous review of the emergence of professional societies, this section describes other global water initiatives, namely the creation of thematic eras after the IHD, the establishment of organized events and the growth of intergovernmental and non-governmental water initiatives.

Designated Periods

The International Hydrological Decade, as noted, was inspired by the International Geophysical Year. Other such time periods have been infrequent, but two are worth mentioning. The first, the International Drinking Water Supply and Sanitation Decade (IDWSSD), was declared in 1981, commencing 6 years after the end of the IHD. This effort aimed to redress massive shortages of access to potable water and sewerage. A dozen years after the IDWSSD ended, it was clear that much of the world continued to lack safe drinking water. Beginning at the 1992 Earth Summit in Rio and through to the 2002 World Summit on Sustainable Development in Johannesburg, experts, officials and activists began calling for comprehensive steps to address this crisis (Cosgrove 1999; Cosgrove and Rijsberman 2000). Improving water management, according to this view, could only be achieved via far-reaching measures that would include population reduction, improved women's education, reformed modes of water governance, and new economic approaches. A recent attempt to realize some of these aims was the 2003 International Year of Freshwater (IYF). It constituted an important effort to improve management practices and to raise public awareness regarding the relationship between water, poverty alleviation and development (Brewster 2004). As with most such initiatives, results have been difficult to measure and some have questioned its effectiveness.

Other such periods have been proposed. The most recent example of such an initiative was in September 2003, when representatives of 53 countries at a global freshwater forum in Tajikistan appealed for a new decade to concentrate on quality-of-life issues relating to water. In response, the UN has created the International Water for Life Decade, which began on World Water Day, 22 March 2005. Table 3.2 lists designated periods of water awareness.

Table 3.2 Designated time periods relating to water awareness

Designated period	Year(s)
International Hydrological Decade (IHD)	1965–1974
International Drinking Water Supply and Sanitation Decade (DWSSD)	1981–1990
International Year of Freshwater (IYF)	2003
International Water for Life Decade	2005–2015

Sources: "Milestones", WWAP (2003b: 24–28); and authors' compilations.

Organized Events

In addition to the designated time periods, another extremely common type of global water initiative has been the organized conference.[6] Both modern diplomats and academics have evinced a fondness for large "watershed" summits that unite diverse participants and aim to resolve outstanding issues. Notable events, many of them sponsored or co-sponsored by UN agencies, at which water was a major topic are shown in Table 3.3.

The periodic events generally have been well attended and have fielded ambitious, wide-ranging and crowded agendas. Usually, these summits have yielded thoughtful, well-intentioned statements, declarations, plans or other documents. But often it seems that the energy and enthusiasm that are manifest at these gatherings dissipate rapidly and leave few lasting traces. Indeed, the elusive outcome termed "networking" may best characterize the benefits of such Forums. In a following section, we examine the responses of water experts who were asked about their views on the influence of such events. Paradoxically, it appears that even as the popularity and legitimacy of such programmes grow, their effectiveness remains relative, unmeasured and not always evident (Salman 2004; Speth 2003; Falkenmark 2001).

Intergovernmental and Non-governmental Water Initiatives

The cauldron of ideas and activity generated by professional societies, the IHD and other designated periods, summits and the IHP clearly elevated the profile of global water issues. A more lasting impact may be that the existence of these institutions spawned new alliances and organizations. At certain times elements of the above institutions came together to pursue distinctive water-centred agendas.

Beginning in the early 1950s, but especially in the years following the IHD – and often prompted or supported by the IHP and other UN agencies such as WMO and the United Nations Environment Programme, or UNEP (which was itself an offspring of a megaconference, the 1972 UN Conference on the Human Environment) – numerous multinational initiatives were launched. Table 3.4 provides a sample of the most prominent initiatives. Some of these were aimed

[6] Some of these conferences have been so large and ambitious in scope that observers have termed them "megaconferences". For the purposes of the present book, the term has been used by A. K. Biswas to apply to "significant" water conferences, regardless of their size and scope.

at particular water-related sectors (e.g. irrigation and agriculture, waterworks construction, water supply and allocation, drinking water and sanitation, public health, inland basins, groundwater, wetlands, ocean waters, climate and ice); some represented disciplinary orientations (e.g. hydrology, ecology, climatology, environmental health, social sciences and law) and some were expressions of particulars visions (e.g. sustainability, food and water security, interdisciplinarity, environmental justice, "environmentology", stakeholder involvement, science-policy dialogues and conflict resolution). Of special note are a set of initiatives intended to improve the acquisition, management, dissemination and use of water-related data. IHP's FRIEND initiative and the WMO Hydrology and Water Research Programme's WHYCOS, for example, both represent important attempts to coordinate the development of high-quality, regional-level data, and facilitate its exchange among scientists and resource managers. In 1999 IHP and WMO jointly formed the HELP initiative to translate the kinds of datasets and professionals networks produced by FRIEND, WHYCOS and similar programmes into real-world management contexts using the integrated water resource management (IWRM) paradigm.

Table 3.3 Significant events relating to water

Designated event	Declaration	Year	Venue
UN Conf. on the Human Envir.	Stockholm Decl.	1972	Stockholm
UN Conf. on Water	MDP Action Plan	1977	Mar del Plata, Arg.
Intl. Conf. on Water and Envir.	Dublin Statement	1992	Dublin
UN Conf. on Envir. and Devel.	Agenda 21	1992	Rio de Janeiro
First World Water Forum	Marrakech Decl.	1997	Marrakech
Intl Conf. on Water and Sust. Devel.	Paris Decl.	1998	Paris
Second World Water Forum	World Water Vision	2000	The Hague
UN Millennium Assembly	Millennium Decl.	2000	New York
Intl Conf. on Freshwater	Ministerial Decl.	2001	Bonn
World Summit on Sust. Devel.	Johannesburg Plan	2002	Johannesburg
Third World Water Forum	Kyoto Minist. Decl.	2003	Kyoto

Sources: "Milestones", WWAP (2003b: 24–28); and authors' compilations.

Some of the largest, most active and arguably best-financed global water initiatives of this type have arisen quite recently, since 1996. Two of these, the World Water Council (WWC) and the Global Water Partnership (GWP), have palpably activist aims and appear to be the most ambitious and comprehensive. They mean to promote particular, forward-looking approaches to water management. While both seek to improve access to water and thus reduce poverty and enhance security, they have adopted different strategies *en route* to achieving this common long-term goal (Delli Priscoli 2005). The WWC has sought to realize this overarching objective by structuring itself as a forum or think tank that brings into articulation a range of existing organizations for the purpose of developing innovative water-management strategies. The GWP was designed to focus specifically on promoting the synchronization of activities among development-related water organizations in an effort to more effectively use the limited resources allocated to the "developing world's" water sector. Two other organizations, the Global International Waters Assessment (GIWA) and the World Water Assessment Programme (WWAP), are less activist; as their names imply, they seek instead to assess the world's water situation. WWAP is distinctive as a rare concerted effort to attempt to systematically aggregate and synthesize knowledge of the world's water resources and their use. It is an example of how a collection of UN bodies and agencies can collaborate on a cross-cutting issue of significance.

The initiatives discussed briefly in this chapter represent a cross-section of the movements, organizations and efforts that have arisen over the past decade. The advent of these internationally oriented, non-governmental and intergovernmental institutions is a development with parallels in other domains such as public health and agriculture. Like those, it is distinctive and remarkable. The networks within which these initiatives function, the connections between organizations, and the varieties of missions and strategies expressed are as yet poorly understood and merit further study.[7]

3.4 Methodology

As noted earlier, most of the information in the previous section was gleaned from archival research and literature reviews. To supplement that information with more personal or "first-hand" observations of international water-resources professionals, two questionnaire surveys were developed and administered to key leaders and experts in an effort to elicit their opinions about the effectiveness and impacts of selected global water initiatives (GWIs). In the following sections, we describe the processes used to define and identify the potential respondents, develop the structure of the survey instruments, and adopt and implement the techniques used to analyse the results.

Before continuing, the rationale for and consequences of the decisions discussed below merit explicit consideration. As defined, the study population represents a particular segment of the remarkably diverse field of actors involved in water management and research. This segment comprises many of the water sector's elite innovators and decision-makers and was selected strategically to gauge the perceptions of persons well positioned to define the shape of the water sector

[7] The authors recognize that the significant role of the World Meteorological Organization in shaping the global water sector remains underdeveloped in this current account. While space does not permit greater elaboration here, this and other omissions will be addressed in future research.

at a global level. Such individuals and their institutions tend to be disproportionately located in the "First" or "Developed" World.

As such, the current study's findings are neither intended nor should they be taken as representative of "the water sector", writ large, but are valuable insofar as they reflect the perspectives of those who have had and/or continue to wield influence in the sector. This is not to suggest that the contributions of local and regional water managers and researchers are less significant or even that the activities of the sector's elite are necessarily over-determining. Instead, the authors believe that the selected population for sampling is but one of many deserving further studies for its unique contributions to the world of water, one which has to date received remarkably little systematic interrogation.

Table 3.4 Examples of influential intergovernmental and non-governmental global water initiatives

Institution	Date established
International Commission on Large Dams (ICOLD)	1928
Intl Commission on Irrigation and Drainage (ICID)	1950
Working Group on Representative and Experimental Basins	1965
Ramsar (Ramsar Convention on Wetlands)	1971
Intl Hydrological Programme (IHP; based at UNESCO)	1975
World Climate Research Programme (WCRP)	1980
Intl Water Management Institute (IWMI; formerly IIMI)	1984
Flow Regimes from Intl. Experimental and Network Data (FRIEND)	1985
GEWEX (Global Energy and Water Cycle Experiment)	1988
Intergovernmental Panel on Climate Change (IPCC)	1988
Water Supply and Sanitation Collaborative Council (WSSCC)	1990
Intl. Human Dimensions Prog. on Global Environmental Change (IHDP)	1990
Biospheric Aspects of the Hydrological Cycle (BAHC)	1992
World Hydrological Cycle Observing System (WHYCOS)	1993
Intl Network on Participatory Irrigation Management (INPIM)	1994
Global Water Partnership (GWP)	1996
World Water Council (WWC)	1996
World Commission on Water for the 21st Century	1998
Global Intl Waters Assessment (GIWA)	1999
Hydrology for Environment, Life and Policy (HELP)	1999
World Water Assessment Programme (WWAP)	2000
Dialogue on Water and Climate/Co-operative Programme on Water and Climate (DWC/CPWP)	2001
Dialogue on Water, Food and the Environment (DWFE)	2001
Global Water System Project (GWSP)	2002
UN Water	2002

Source: Authors' compilation.

3.4.1 Sampling and Data Acquisition

After careful consideration, the population most appropriate for the study's parameters was determined to be one defined as individuals who significantly have helped shape the development and activities of GWIs. With that as a defining criterion, two types of potential respondents were identified: (1) *representatives* who are high-level professionals and administrators, working for or directly involved with particular GWIs and (2) *observers* who, though not directly affiliated with a particular initiative, are nonetheless highly knowledgeable or experienced about GWIs.

The sample of representatives was drawn from a population of current chairpersons, executive directors, secretaries general, or persons in similar leadership positions affiliated with 38 influential organizations in the fields of water research and management. They were instructed to respond to the survey as individuals, but in reference to their specific organization.[8] A sample of observers was drawn from a population identified through surveys of the scientific literature, reviews of lists of former GWI officials, and via "snowballing" (i.e. by asking colleagues, contacts and other informants to recommend other potential respondents). As experts, the observers were asked to respond as individuals, but with reference to the wider field of international water research and management.[9]

Using these methods, the authors identified 54 representatives and 63 observers to survey (Table 3.5), a total of 117 individuals (91 males and 26 females). The individuals were contacted during the spring of 2004 and were asked to participate in the survey. Response rates for both groups surveyed, representatives and observers, was relatively high, approaching 63 and 60%, respectively (Table 3.5). Females responded at a significantly lower rate (27%) than males (69%). Although the number of women who were invited to participate (and then opted to do so) was significantly lower than that of men because of their relatively small numbers within the particular age and position cohorts sampled for this study (as judged from both literature reviews and snowball sampling) the ratio of women to men in the final sample can be considered reasonably representative of their distribution within the particular population of interest.[10] In general, the authors sought to achieve a balanced sample composition, especially with regard to the number of representatives versus observers, gender of respondents and diversity of institutional affiliation.

---

[8] The 38 influential organizations that were identified for inclusion in the study were selected based on reviews of the literature and discussions.

[9] That is to say that the basis of selection was "institutional significance" and "individual reputation", respectively, for representatives and observers. Clearly, however, representatives can be considered knowledgeable and respected individuals in their own right.

[10] Considerable deliberate effort was made to enrol female observers and representatives in the study. As a result, the percentage of females invited to participate exceeds the proportion of those active in the field. Via self-selection, however, the eventual proportion of female respondents was considerably lower.

Table 3.5 Survey respondent types

	Identified	Completed at least one survey	Response rate (%)
Representatives	54	34	63
Observers	63	38	60
Males	91	63	69
Females	26	7	27

Source: Authors' compilation.

The 117 persons identified by the authors were contacted and provided copies of the research instruments via e-mail, fax, postal service or in person. Of those, 82 agreed to participate (Table 3.6). Subsequent communications between Varady and potential respondents continued throughout 2004 in an effort to encourage participation. Only nine of the individuals contacted (8%) declined to participate, while about a quarter (22%) did not respond.

To date, 70 individuals (60% of those invited, or 85% of those who agreed to participate) have responded to one or both of the surveys (usually in writing but in several cases verbally; 16 individuals granted interviews or communicated more informally with Varady regarding one or more of the study's central themes). Approximately the same number of persons completed each of the two forms (54 for the long form, 57 for the short form), both of which are above the minimum needed (44) to obtain results valid with a 90% confidence interval.[11]

Table 3.6 Surveys response summary

Action	Number of individuals	Percentage of initial survey pool	Percentage of those agreeing to participate
Contacted	**117**	100	
Declined to participate	9	8	
No response, dropped from list	26	22	
Agreed to participate	**82**	**70**	100
Completed one or both surveys*	70	60	85
Long form and interview	54	46	66
Short form (influences)	57	48	68

Source: Authors' compilation.

* See next subsection for a description of the two principal instruments.

3.4.2 Data Acquisition, Management and Analysis

Based on the types of information we hoped to obtain from the respondents, two types of questionnaires were deemed necessary. The first survey instrument was designed to gauge the relative "influence" of selected GWIs. The instrument was

[11] The sample size (n) required for a confidence interval of $\alpha = 0.10$ with a standard deviation of 1.33 and a tolerable error of $_{(+/-)}$ 1/3rd point is 44 ($n = [(Z_{\alpha/2})^2\sigma^2]/E^2 \approx 43.34$).

built around a Likert scale and was intended to produce quantitative measures of perceived influence that would facilitate a more objective comparison of the 30 GWIs that were evaluated using the survey. The second instrument, an open-ended questionnaire, was structured to obtain qualitative responses to a number of questions related to intellectual and practical issues; organization background; governance; successes and failures; institutional overlap, cooperation and competition; and general trends and assessments.

Survey 1: Short Form

The first survey instrument (Appendix A), referred to as "Survey 1" or "short form", was designed to elicit respondents' evaluations of the relative influence of 30 different GWIs,[12] both from the perspective of particular organizations of interest and relative to the field of global water research, policy and management more generally (via representatives[13] and observers, respectively).

In the survey these 30 GWIs were grouped into three general categories: (1) professional societies, (2) designated time periods and organized events and (3) intergovernmental and non-governmental organizations. Within each category, GWIs were listed chronologically (in the cases of events and time periods) or alphabetically (in the case of societies and organizations).

Respondents were asked to rate the intensity of the influence of each GWI by assigning to it a single integer value from a five-point Likert scale[14] in which intensity of influence ranged from "very strong" to "very weak or non-existent" (from one to five, respectively).[15] Responses, along with variables indicating respondents' gender and observer/representative status, were compiled by the authors in a spreadsheet for basic descriptive analysis. The dataset was also imported into the statistical analysis package, SPSS.[16] Comparison of means between (a) observers and representatives and (b) women and men were conducted to determine whether statistically significant differences of means could be detected along either of these dimensions. In the case of observers/representatives, the results of a *t*-test[17] yielded statistically significant differences for 6[18] of the 30 GWIs,

[12] Of the 56 GWIs listed in Tables 1.1 through 1.4, 30 of the most active and prominent were selected for this survey.

[13] The design of the instrument did not permit representatives to rate the relative influence of the GWIs with which they were affiliated (and in any event all self-evaluations were excluded from the analysis).

[14] See Likert (1932) for one of the first explications of this approach.

[15] See De Vellis (1991) and Spector (1992) for discussions of theory and issues behind scale construction and Bernard (1995), Calder (1996), Healy (1990) and Marsh (1988) for discussions of the use of scales in social science research.

[16] SPSS Release 11.5.0, SPSS Inc.

[17] In this test α was set at 0.05 in a two-tailed *t*-test adjusted (where appropriate) for unequal variances.

[18] These exceptions were the International Hydrological Decade, the UN Conference on the Human Environment, the International Drinking Water Supply and Sanitation Decade,

suggesting that, in general, differences between the two groups were relatively minor. In the case of gender, no statistically significant differences were discernable between men and women.[19] Consequently, in this chapter the authors do not distinguish between categories *a* and *b* when presenting and discussing results.

Survey 2: Long Form

The second instrument (Appendix B), also referred to as "Survey 2" or "long form", was intended to complement the ratings of influence adduced from the short form by providing participants with the opportunity to express more expansively their perceptions regarding a range of related issues. For example, Survey 2 included questions regarding the constitution and relative significance of different ideas, practices and socio-political forces in helping to shape the evolution of GWIs.

Further, the instrument sought sophisticated descriptions of the activities of and relations between different actors and organizations, as well as detailed explanations of the development of problems faced, opportunities encountered and strategies employed in the context of these articulations. Finally, informants were asked to elaborate on how the field is currently developing and to offer prescriptions for where it should be going.

To obtain such accounts and explanations, a 40-question, open-ended response survey was developed and formulated around the following thematic domains: (1) organizational background, (2) intellectual currents, (3) socio-political currents, (4) issues and practices concerning governance, (5) evaluation of initiatives' success or failure, (6) inter-institutional relations and (7) perceptions and evaluations of general trends.

Individual questions within each of these categories were developed in the light of primary- and secondary-source literatures and Varady's own preliminary research. Additionally, these benefited significantly from the suggestions of a number of generous collaborators and colleagues. Through this process of review and revision, questions were developed that not only addressed various aspects of the thematic domain in which they were embedded, but did so in a way that systematically linked these to the more general research interests outlined in the beginning of the chapter.

Although structured at this level, because the questions were left open-ended, respondents were allowed to augment, elaborate and even propose alternative themes, meanings and orientations that the authors could not have, in some cases, even envisioned, much less accommodated in a more structured format. At the same time, we note that, naturally, not all informants responded to all questions; indeed, their invitations made clear that they could concentrate on those questions they felt were most relevant to their experiences.

UN Conference on the Environment and Development, the International Year of Freshwater and the Third World Water Forum.

19 It should be pointed out, however, that the small sample size of the women's cohort ($nw = 5$) could render any real population-level difference imperceptible.

At first glance, such a format would appear to make comparative analysis more complicated (relative to, e.g., the short form). However, this disadvantage is more than offset by the instrument's capacity to (1) grant the researcher the ability to address multiple and complex issues, (2) permit the respondent far more flexibility in responding to questions and, ultimately, (3) generate data sets characterized by far greater multi-dimensionality and nuance than would be possible with more structured and more easily-managed instruments.

Analysis of the responses to Survey 2 was inherently complex. For the purposes of this chapter, the authors did not employ a discursive analysis but decided instead to approach the data using more straightforward survey interpretation techniques. Specifically, the authors imported the survey data into a popular qualitative data-management and analysis package, called N6,[20] to facilitate the categorization, disaggregation, and sorting of the data set. Through a series of extensive coding procedures, textual data was transformed into categorical data that could be expressed as frequencies and presented in the form of tabulations capable of offering accessible depictions of the distribution of different response types.

For example, with a survey question such as "What is the role of governments in the work of global water initiatives?" the above technique allows the authors to identify and assign responses into such niches as "offering leadership", "providing funding", "advising", "enhancing participation", "facilitating stakeholder involvement", and so on. Once the responses are thus grouped, they can be ordered by frequency and then interpreted.

For the purposes of this chapter, of the 40 questions in Survey 2, eight were selected for analysis based on their centrality to the paper's hypothesis and with particular attention to the thematic foci of this conference. Paraphrased and grouped, the questions are as follows:[21]

- What roles do different constituencies, governments, non-governmental organizations and stakeholders, play in the work of GWIs? (Table 3.11)
- What actions have GWIs engaged in and what have these accomplished? (Tables 3.12 and 3.13)
- How do respondents view the overlap and proliferation of GWIs? (Tables 3.14, 3.15, and 3.16)
- What, overall, has been the significance of GWIs for the global water sector? (Table 3.17)
- What lessons have been learned in the process? (Table 3.18)

[20] N6 is a recent version of NUD*IST and a product of QSR International Pty Ltd.

[21] Note: the actual survey questions are included (verbatim) under the title of each figure.

3.5 Survey Findings and Interpretation

3.5.1 Survey No. 1: Measuring the Relative Influence of Individual Global Water Initiatives

As described in the methodology section above, the short-form survey instrument was designed to obtain numerical ratings of the perceived “influence” of 30 different initiatives of four types. The number of responses registered, 56, was sufficiently large in all cases to yield meaningful and statistically valid ratings.[22] In this section, we present the results of the responses to Survey No. 1. The authors begin by considering the overall distribution of scores and the ratings of GWI categories and conclude with the presentation and discussion of the survey results for individual initiatives (see Tables 3.7, 3.8, 3.9 and 3.10.)

General Observations

- Figure 3.2 shows the frequency of assigned ratings for four aggregated categories of GWIs (professional societies, designated periods, events and organizations). As that figure indicates, when taken together GWI scores (1) are nearly evenly distributed around a rating of three, (2) exhibit a relatively flat profile (kurtosis = –1.10), and (3) are only slightly skewed (skew = –0.02) towards ratings of lesser influence. Individually, the distribution of scores for the category “Organizations” most closely follows the collective trend while that for “Events” is skewed towards higher perceived influence.
- The composite mean score for all 30 GWIs was 3.03, with category means ranging from a best of 2.84 (lower is better) to a worst of 3.19. Translating the “1” (best) to “5” (worst) ratings to a grade scale of “A” to “F” (standard for the United States higher educational system), 3.03 can be interpreted as a “C,” or mediocre, grade. In other words, respondents were not inclined to rate the influence of the initiatives highly. This suggests that as a group, the respondents were rather sceptical.
- Further evidence of cynicism is rendered legible by Figure 3.3: only 12 of the 30 initiatives received a rating higher than C; the highest rating assigned was B^- (2.0–2.5), which was awarded to only five GWIs, while four were assigned a poor, D^+ rating (3.5–4.0).
- Among the four categories (professional societies, designated time periods, organized events and intergovernmental and non-governmental organizations), events were considered the most influential (with a rating of 2.84) and organizations were perceived as least influential (with a rating of 3.19).

[22] Of course, not all respondents rated every GWI, but in no case did the sample size fall below 44, the minimum needed to achieve 90% confidence.

- Finally, a test to determine strength of association between a GWI's date of origination and its mean rating showed no statistically significant correlation between the two variables.[23] This is interesting because one might have hypothesized that, for example, older GWIs would be advantaged relative to younger ones on account of having had more time to exert influence or, alternatively, that younger initiatives could be perceived to exert greater influence on account of their immediate relevance and prominence in contemporary discourse. A demonstration of no correlation between date of origination and mean rating indicates, minimally, that if such factors are at all effective, they are not over-determining.

3.5.2 Results and Findings for Individual Initiatives and by GWI Category

In the following subsection, survey results (including mean scores, standard deviations and sample sizes) are grouped by GWI category and presented in tabular form for the study's 30 initiatives. Each of these four tables is followed by brief summaries of the principal findings.

Observations on Professional Societies (Table 3.7)

- The society receiving the highest rating, that is the one considered the most influential, is also one of the oldest (established in 1922) and most venerable: the International Association of Hydrological Sciences (IAHS). The 4000-member association's mission has been simple and essentially unchanging: *to promote the study of hydrology*. With a rating of 2.30, IAHS was the only professional society to score below 3.0 (Rodda 1999; Volker and Colenbrander 1995).
- The next highest-rated institution was a sub-specialty society, the International Association of Hydrogeologists (IAH), which has 3500 members and was established in 1956 to promote research, "proper" management, and protection of groundwater (Day 1999). IAH was rated at 3.04, or C.
- The International Water Association (IWA), founded in 1999, is a much newer society with a very broad mission encompassing both research and practice. IWA, which has made a mark in publishing respected books and journals, fared nearly as well as IAHR among respondents, securing a 3.08 rating.
- The International Water Associations Liaison Committee (IWALC) is an even more recent creation. IWALC was intended to promote communication and working relationships among leaders of 10 professional societies with differing aims. Its low profile and relative inactivity are reflected in its low rating of 3.79.

---

[23] With α set at 0.05.

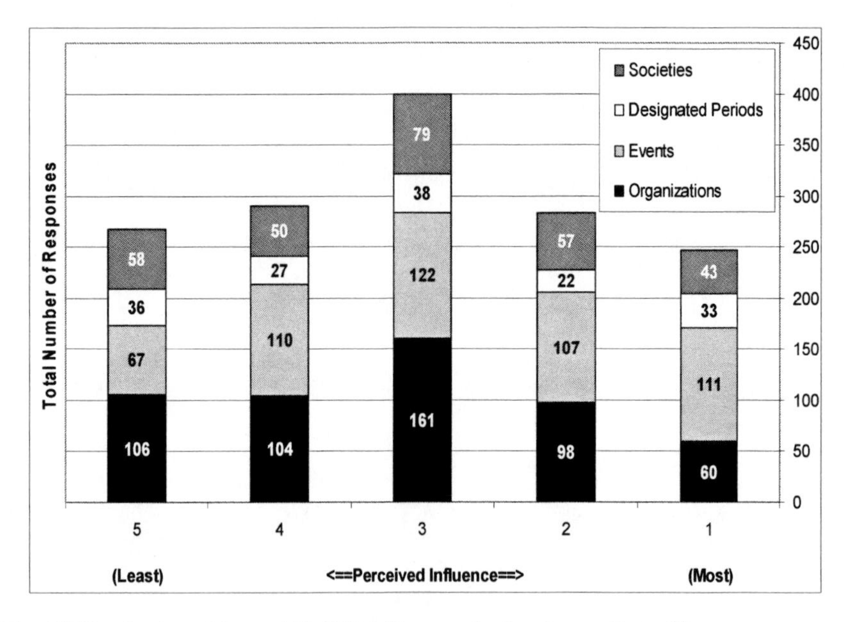

Fig. 3.2 Distribution of Score (all GWIs) (Source: Authors' compilation[24])

Table 3.7 Perceived influence of professional societies

Professional societies	Mean Rating	Standard Deviation	*N*
International Association of Hydrological Sciences (IAHS)	2.30	1.27	50
International Association of Hydrogeologists (IAH)	3.04	1.46	49
International Water Association (IWA)	3.08	1.24	50
International Water Resources Association (IWRA)	3.15	1.15	48
IAHR (formerly International Association for Hydraulic Research)	3.22	1.21	46
International Water Associations Liaison Committee (IWALC)	3.79	1.27	45
All professional societies*	3.08	1.33	288

Source: Authors' compilation.

* The mean ratings and standard deviations associated with the row titled "All Professional Societies" were calculated using all scores for all Societies in aggregate. The same is true for the last row in Tables 1.7b–d as well.

[24] Numbers in columns represent the total number of responses offered by the entire pool of respondents per score per category of GWI. For example, a rating of "5" (i.e. "very weak to non-existent" influence) was given 106 times to the GWIs constituting the category "organizations" by the 57 respondents.

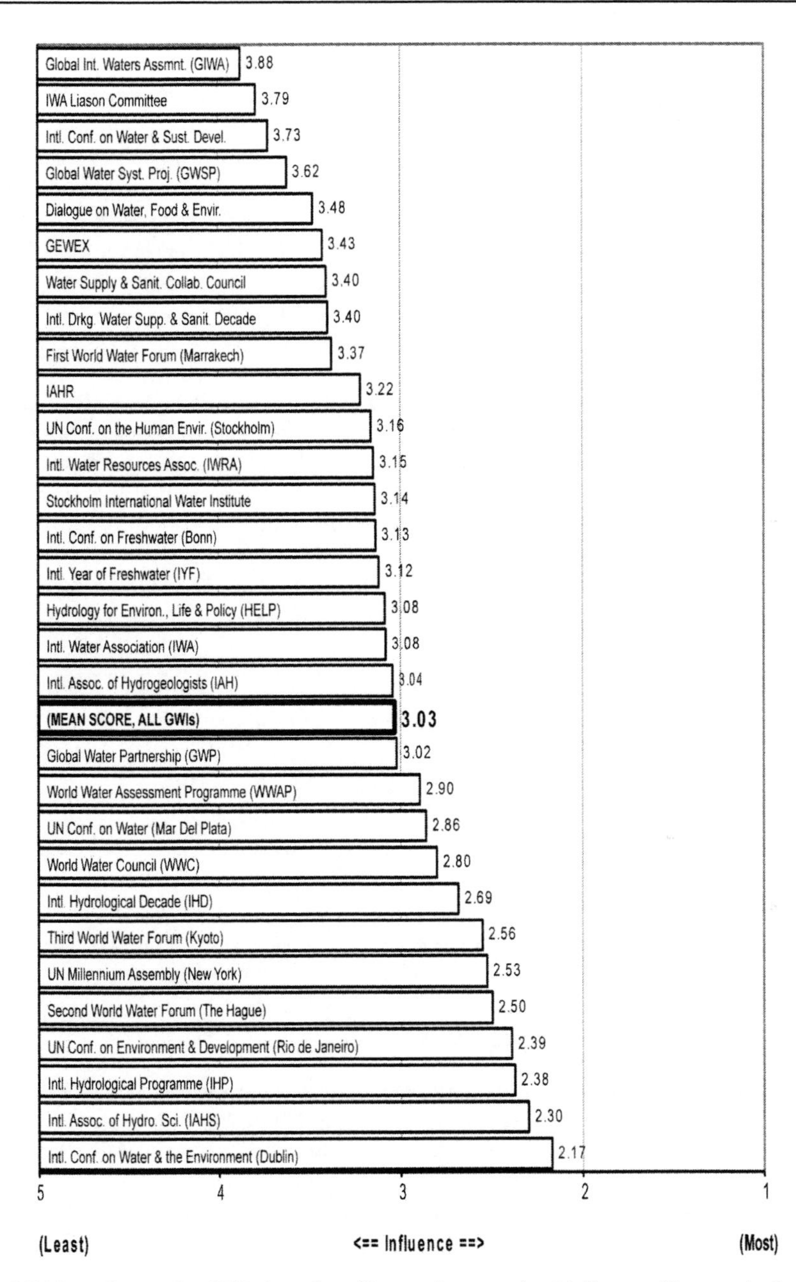

Fig. 3.3 Mean Scores for GWIs in order of increasing perceived influence (Source: Authors' compilation)

Observations on Designated Time Periods (Table 3.8)

- The International Hydrological Decade (1965–1974), rated at 2.69, was considered far and away the most influential time period. As a recognized catalyst for many subsequent developments in hydrological science, education, training and implementation, the IHD is seen by many long-time observers as the model *par excellence* for raising awareness by means of designating time periods (see, e.g., Korzoun (1991) and Entekhabi et al. (1999)). New initiatives, such as the 2003 movement to create the current International Water for Life Decade (2005–2015) uniformly draw their inspiration from the IHD.
- At the other end of the spectrum, the International Drinking Water Supply and Sanitation Decade (IDWSSD) was poorly regarded, receiving a score of 3.40. The IDWSSD, which was in place during the 1980s (1981–1990), unlike other such efforts which sought mainly to enhance science and understanding, was the first such period to target real, on-the-ground improvement in conditions. That many of its goals were unrealized was a disappointment to many observers, who accordingly saw it as not very influential.

Table 3.8 Perceived influence of designated time periods

Designated time periods	Mean Rating	Standard Deviation	*N*
International Hydrological Decade*	2.69	1.66	51
International Year of Freshwater*	3.12	1.37	52
International Drinking Water Supply and Sanitation Decade (DWSSD)*	3.40	1.23	53
All time periods	3.07	1.45	156

* Observer and representative ratings showed statistically significant difference for these GWIs. (Source: Authors' compilation).

IHD (Obs. = 2.09; Rep. = 3.18); IYF (Obs. = 3.56; Rep. = 2.70); IDWSSD (Obs. = 3.04; Rep. = 3.71).

Observations on Organized Events (Table 3.9)

- Of particular relevance to the present text, it is noteworthy that of the 10 events surveyed, with a 2.17 score, the International Conference on Water and the Environment, commonly referred to as the Dublin Conference (1992) drew the highest rating, not only among peer events, but among all GWIs. This meeting was just the second of its kind devoted to the subject of water (Mar del Plata was the first) and it was designed to set the water-related agenda for the Rio Conference (officially, the UN Conference on Environment and Development, or Earth Summit) held later that year. Some 500 persons from 100 nations attended and drew up the Dublin Statement, which has come to be recognized as a seminal document that has influenced all subsequent declarations (The Dublin

Statement on Water and Sustainable Development 1992; UNACC and ISGWR 1992; Salman 2003, 2004).

- The Earth Summit, although a much larger and less focused gathering, also was deemed influential by the survey respondents, drawing the next highest rating, 2.39, and demonstrating that compactness is not a determinant of success for such meetings.
- So far, there have been three World Water Forums since 1997, with a fourth scheduled for 2006. The first of these, held in Marrakech (WWF1), drew a modest audience and was not viewed as particularly influential by the respondents, who assigned a score of 3.37, ranking WWF1 just ahead of the Johannesburg Summit of 2002, in the next-to-last place. The much larger Second World Water Forum (WWF2), in the Hague, The Netherlands, in 2000, was considered the most influential of the three, scoring a respectable 2.50. WWF3, the 2003 Kyoto event, the largest of the three, had a slightly lower, 2.56, rating. Among participants and observers, the three Forums have drawn mixed reviews. The most frequent criticisms of WWF3 have been the event's vastness, its all-inclusiveness and the concomitant diffusiveness of the programme, its inability to articulate a coherent and forceful declaration and, finally, among certain community groups and non-governmental organizations, fears of its unstated intentions regarding privatization of water (Kuylenstierna 2003).
- The 1977 Mar del Plata Conference (formally, the UN Conference on Water) was the first major conference to deal exclusively with the subject of water. Its Action Plan recognized the need to address water issues at all levels, national, regional and international. The principles embodied in the plan set the stage for future declarations, pronouncements and approaches. Yet nearly three decades after Mar del Plata, most observers agree that many of that conference's recommendations remain to be fulfilled. For that reason, perhaps, or maybe because of the event's distance in time, it was not rated as particularly influential, receiving a score of just 2.86, placing the event sixth among the 10 events considered (Najlis and Kuylenstierna 1997).
- The 2002 Johannesburg World Summit on Sustainable Development ranked last in this category. Coming 10 years after Rio, which as we saw was highly regarded, Johannesburg generally disappointed attendees and experts, especially those in the field of water. The Summit's theme was deliberately steered towards development, as opposed to environment, and this set a tone for those with interests in ecology, conservation and better access to water and sanitation. Additionally, the high, perhaps unprecedented degree of politicization that preceded the Summit further impeded its success. Unsurprisingly, therefore, it scored last with 3.73 (Speth 2003).

Table 3.9 Perceived influence of organized events

Organized events	Mean Rating	Standard Deviation	*N*
International Conference on Water and the Environment (Dublin, 1992)	2.17	1.26	52
UN Conference on Environment and Development (Rio, 1992)*	2.39	1.25	51
Second World Water Forum (the Hague, 2000)	2.50	1.22	54
Millennium Development Goals/UN Millennium Assembly (NYC, 2000)	2.53	1.33	51
Third World Water Forum (Kyoto, 2003)*	2.56	1.21	54
UN Conference on Water (Mar Del Plata, 1977)	2.86	1.31	51
International Conference on Freshwater (Bonn, 2001)	3.13	1.22	52
UN Conference on the Human Environment (Stockholm, 1972)*	3.16	1.39	50
First World Water Forum (Marrakech, 1997)	3.37	1.23	51
World Summit on Sustainable Development (Johannesburg, 2002)	3.73	1.13	51
All events	2.84	1.33	517

* Observer and representative ratings showed statistically significant difference for these GWIs. (Source: Authors' compilation).

UNCED (Obs. = 2.04; Rep. = 2.73); TWWF (Obs. = 3.08; Rep. = 2.10); UNCHE (Obs. = 2.63; Rep. = 3.65).

Observations on Intergovernmental and Non-governmental Organizations (Table 3.10)

- Of the 10 organizations rated by respondents, UNESCO's International Hydrological Programme (IHP) received the highest score, 2.38, nearly a half-point higher than the next most highly rated initiative. This affirmation of IHP's centrality and influence is not surprising in view of several factors: (a) its venerability (IHP recently celebrated its 30-year anniversary) and quasi-permanence;[25] (b) its situation as part of the UN System, within UNESCO, the agency with the largest of the many UN programmes on water; (c) the direct participation of more than 160 national governments in setting IHP's priorities and the legitimacy this association confers; (d) the availability of budgetary resources stemming from this arrangement; and (e) its geographic as well as institutional centrality, which encourages frequent face-to-face meetings, discussions, and joint enterprises.
- Two of the next three highest ratings were assigned to the World Water Council (WWC), which earned a 2.80 score, and the Global Water Partnership (GWP),

[25] IHP functions in 6-year cycles, through which the organization is renewed and its priorities and programmes adjusted and reoriented. The current cycle is IHP-VI (2002-07).

which received a somewhat lower 3.02. That these two institutions are grouped near the top is of special interest for a number of reasons. They were established the same year, in 1996. They also were created more or less in concert and received initial support from some of the same donor agencies (notably, the World Bank and the UN Development Programme). By design, they were intended to be opposite sides of a coin. Thus, WWC was conceived as a think tank that would "promote awareness, build political commitment and trigger action", while GWP was seen as an on-the-ground, action-oriented organization that would help set up functional partnerships among all those involved in water management. The Council operates at a secretariat in Marseille, France (the host city of Marseille provides WWC's offices and much of its staff) on behalf of some 330 dues-paying institutional members. The Partnership is based in Stockholm, Sweden, and receives fund annually from an assortment of donors as well as from the Swedish government. While publicly the two initiatives profess to collaborate and to address non-overlapping issues, there are nevertheless palpable undertones of competition between supporters of WWC and GWP, if not between the two initiatives themselves. In reviewing the responses provided by survey informants, there were numerous instances of inverse scoring patterns for the two organizations, that is high scores for one were accompanied by low scores for the other and vice versa. To the extent that this survey offers any indication of experts' opinions on relative influence, WWC has achieved the higher rating.

- Like WWC and GWP, two other water initiatives, also ranking in the top five, can be seen to be "friendly" rivals. The World Water Assessment Programme (WWAP) and the Hydrology for Environment, Life and Policy (HELP) initiative both are hosted by UNESCO,[26] and as such, compete to some degree for funds and influence within IHP. But there the similarities end. WWAP, by far the larger of the two enterprises, has been primarily funded by external sources, to date, the Japanese, Spanish and Italian governments have been the most significant. Further, WWAP, which was created in 2000, derives its mandate from the United Nations at-large, being a creation of an umbrella group called UN Water, a collection of the leading water-related programmes within the UN family. WWAP's primary charge has been to conduct periodic assessments of the world's water resources and conditions, and to prepare triennial reports on its findings (WWAP 2003a; WWAP 2005).[27] By contrast, HELP, since its 1999 establishment, has been supported mostly by IHP funds, at a much lower level than WWAP, and was officially included within IHP-VI, the IHP's 6-year workplan. Unlike WWAP, it does not have a mandatory charge and its mission is more dilute. In its own words, HELP seeks to develop "a new approach to integrated catchment management through the creation of a framework for water

[26] HELP is officially a joint undertaking of UNESCO and its sister agency, the World Meteorological Organization (WMO). See HELP (2005).

[27] The reports are meant to be issued on the occasion of the World Water Forums. Thus, the first World Water Development Report, prepared by WWAP, appeared at the 2003 Kyoto Third World Water Forum.

law and policy experts, water resource managers and water scientists to work together on water-related problems."[28] In short, HELP is establishing a network of some 70 basins around the world, each of which subscribes to a recommended approach to water management. In practice, only one of the HELP basins has served as a case study for WWAP. Nor have there been other notable overlaps between these two initiatives. Nonetheless, the two initiatives would seem to be competing, if not over similar agendas, then for attention, recognition, legitimacy and ultimately, influence. WWAP, which was rated at 2.90, scored somewhat higher than HELP, which garnered a 3.08 rating. The difference is not large, however, suggesting that in spite of their differing missions and disparate resources, many observers have difficulty delineating their relative influence (HELP 2005; HELP Task Force 2001).

- Like WWAP, and as its title indicates, the Global International Waters Assessment (GIWA), is a global assessment programme. Created in 1999 and based in Kalmar, Sweden, GIWA was designed as a fixed-term project, slated to conclude its work at the end of 2003. Its aim is "to produce a comprehensive and integrated global assessment of international waters, the ecological status of and the causes of environmental problems in 66 water areas in the world, and focus on the key issues and problems facing the aquatic environment in transboundary waters"[29] (GIWA 2005). Also like WWAP, GIWA received substantial financial support when it began, much of it from the United Nations Environment Programme (UNEP), the Global Environmental Facility (GEF) and the Swedish government. Unlike WWAP, whose task has been to produce a single report every 3 years, GIWA was expected to issue periodic area reports on the 66 regions selected, as those areas were assessed. Many respondents apparently felt that GIWA had not fulfilled its promise and ranked the institution last among the 10 initiatives rated, giving GIWA a score of just 3.88, or just above D. In March 2005 GIWA distributed a performance survey to its constituents. But other than that activity and a flurry of regional reports issued at the end of 2004, the degree to which GIWA continues to function is unclear.
- Three of the organizations appearing in Table 3.10, the Global Energy and Water Cycle Experiment (GEWEX), the Dialogue on Water, Food and Environment (DWFE), and the Global Water System Project (GWSP) were ranked among the bottom four, receiving scores of 3.43, 3.48 and 3.62, respectively. That these initiatives got relatively poor grades is perhaps attributable to their low profiles and specialized niches. Two, GEWEX and GWSP, are highly scientific and therefore likely unknown to many of the respondents, even though GEWEX has existed since 1988 (GWSP is less than 3 years old). These results yield a useful caveat: *That despite their statistical validity, the ratings provided by the survey respondents obviously include a strong* "PR" *(public relations) component.* As elsewhere, it is not only what you achieve that matters, but how well you inform your public.

---

[28] See the HELP Web site at <http://portal.unesco.org/sc_nat/ev.php?URL_ID=1205&URL_DO=DO_TOPIC&URL_SECTION=201>.

[29] See the GIWA Web site at <http://www.giwa.net/giwafact/giwa_in_brief.phtml>.

Table 3.10 Perceived influence of organizations

Intergovernmental and non-governmental organizations	Mean Rating	Standard Deviation	*N*
UNESCO International Hydrological Programme (IHP)	2.38	1.18	48
World Water Council (WWC)	2.80	1.27	51
World Water Assessment Programme (WWAP)	2.90	1.21	48
Global Water Partnership (GWP)	3.02	1.30	50
Hydrology for the Environment, Life and Policy (HELP)	3.08	1.25	48
Stockholm International Water Institute (SIWI)	3.14	1.17	51
Water Supply and Sanitation Collaborative Council (WSSCC)	3.40	1.26	47
Global Energy and Water Cycle Experiment (GEWEX)	3.43	1.30	47
Dialogue on Water, Food and Environment (DWFE)	3.48	1.07	46
Global Water System Project (GWSP)	3.62	1.21	45
Global International Waters Assessment (GIWA)	3.88	1.10	48
All organizations	3.19	1.27	529

Source: Authors' compilation.

3.5.3 Survey No. 2: Evaluating Views and Opinions of Global Water Initiatives

Results from the long-form analysis appear in the following pages as tabulations (Tables 3.9, 3.10, 3.11, 3.12, 3.13, 3.14, 3.15, and 3.16) showing the distribution of responses for each of the eight questions analysed. In each case, the actual question posed in the survey form appears in quotation marks below the table heading. Significant observations on the response sets to each question appear just below the table corresponding to that question.[30]

[30] However, as explained in the earlier discussion of *Methods*, the response categories used below were derived through the authors' interrogations of data for individual questions (i.e. were not determined *a priori* or developed as part of a master coding scheme) and reflect the specific kinds and combinations of descriptive, thematic, comparative or evaluative responses offered by study participants. (Note: Like-termed categories share an identical definition.)

Role of Government, NGOs and Stakeholders: Observations on Table 3.11

> The lesson learned most widely is that water assistance should be directed to the grass-roots, where spending small amounts will benefit most of those in need.
>
> – John Rodda, Former President, IAHS

- There is overwhelming support (95.5%) for governmental involvement in the work of GWIs.
- Non-governmental organizations (NGOs) and stakeholder groups are similarly considered important players, with 88 and 93% of respondents affirming these roles, respectively.
- For governments, leadership is seen as their greatest potential contribution (34%), while the prospect of obtaining funding[31] (26%) and facilitating broader participation are also viewed as significant.
- Respondents clearly do not expect NGOs and stakeholder groups to provide funding (just 3 and 5%), but they do see both as enhancing participation (50 and 64%, respectively). In the process, stakeholder groups, especially, are seen as having the role of asserting their own agendas (18%).
- Had this question been asked 25 years ago, the answers likely would have been quite different, with much less acknowledgement of the importance of informal, "bottom-up" constituencies, or, as one respondent commented, over the years, "The lesson learned most widely is that water assistance should be directed to the grassroots, where spending small amounts will benefit most those in need" (John Rodda).

Table 3.11 Role of Government, NGOs and stakeholders

Categories of Responses	Government 23 responses 57 responses	NGOs 23 persons 56 responses	Stakeholders 21 persons 36 responses
Functional ($n = 89$)	100% ($n = 35$)	100% ($n = 32$)	100% ($n = 22$)
Offering leadership (legitimacy and authority)	34 (12)	13 (4)	–
Providing funding	26 (9)	3 (1)	5 (1)
Enhancing participation	23 (8)	50 (16)	64 (14)
Advising	11 (4)	22 (7)	9 (2)
Enhancing Flexibility	3 (1)	3 (1)	5 (1)
Asserting own agenda	3 (1)	9 (3)	18 (4)
Evaluative ($n = 60$)	100% ($n = 22$)	100% ($n = 24$)	100% ($n = 14$)
Beneficial ($n = 55$)	95 (21)	88 (21)	93 (13)
Detrimental ($n = 5$)	5 (1)	12 (3)	7 (1)

Evaluation of a sample of 41 of the 52 returned long forms (Survey 2), 14 January 2005.

[31] Commenting on the relationship between governments, funding and GWIs, one respondent claimed that, "in practical terms, the influence of the programmes and initiatives I am aware of has been small on the host bodies and on those at the governmental level participating in them. The main reason for this is the lack of finance to promote these initiatives" (J. Rodda).

Significant Actions: Observations on Table 3.12

GWIs provide a mechanism for cooperation and sharing of ideas, resources and techniques.

– Alan Hall, IAHS/WMO Working Group on GEWEX

The boldest initiatives [promoted] a move toward water user associations, which also had direct positive effect on getting women involved in decision-making and management.

– Eugene Stakhiv, Institute for Water Resources, US Army Corps of Engineers

GWIs are effective in shaping policy agendas because they impose a mainstream blueprint thinking that pervades many agendas. Whether these ideas are good, and to what extent they are put in practice is another story.

– François Molle, International Irrigation Management Institute

Real policy follows capital. If you want to know what's really happening in water policy, don't look to intellectual efforts, but go to the working guidelines for development banks.

– Aaron Wolf, Professor, Oregon State University

- Efforts by GWIs, according to the respondents, have been oriented most prominently to advancing ideas and practices (49%). While 12 responses indicated that GWIs had aided in creating institutional infrastructure, other forms of infrastructure were infrequently cited, indicating perhaps that these may be seen as beyond the capacity of most GWIs. At the other end, promoting cooperation (18%) may be considered not ambitious enough.
- Among the ideas and practices that were thought significant, publications were cited by nearly a third of the respondents (11 of 36), with actual projects and programmes a close second (10 of 36).
- Especially germane to the topic of the Bangkok workshop, of the 74 responses considered, only four (5%) cited the holding and facilitating of meetings and conferences as significant.
- It must be noted that the most frequent response to the question was "unspecified", which in all was offered 16 times. This likely reflects the vagueness of respondents' expectations of GWIs. Thus, even among the 36 answers confirming that advancing ideas and practices were important contributions, 14 (39%) did not articulate particular ideas or practices.

Table 3.12 Significant actions

Categories of responses 32 persons 74 responses	Responses [% (number)]
Advancing ideas and practices	49% (36)
Publications	15 (11)
Projects and programmes	14 (10)
Meetings and conferences	1 (1)
Unspecified	19 (14)

Table 3.12 (Continued)

Categories of responses 32 persons 74 responses	Responses [% (number)]
Developing infrastructure	33% (25)
Institutional	16 (12)
Guidelines, models and toolkits	7 (5)
Research and data	5 (4)
Policies	3 (2)
Technology	1 (1)
"Too early to tell"	1 (1)
Promoting collaboration and cooperation	18% (13)
Developing networks	7 (5)
Holding and facilitating international meetings	5 (4)
Collaborating around joint proposals	3 (2)
Unspecified	3 (2)

Evaluation of a sample of 41 of the 52 returned long forms (Survey 2). 14 January 2005.

Programme Results: Observations on Table 3.13

Few if any of the initiatives have actually been tested against measures of success and where this has been attempted the results have not met the stated objectives.

– John Rodda

Whereas Table 3.12 sought to identify which actions were considered significant, the present question asks which actions have been the most enduring, and by implication, most positive.

- Meetings and conference, poorly regarded in the previous question, once again evinced weak support; only 3 of 77 responses indicated that such events were of lasting value.
- As in Table 3.12, advancing ideas and practices was considered meaningful, with just under half of the responses (43%) indicating this. Also as above, publications were the favoured output (11citations).
- By and large, the trends shown in Table 3.13 parallel those in Table 3.12. This is consistent with the correspondence between action (Table 3.12) and results (Table 3.13).
- However, while only two of the Table 3.12 respondents (3%) thought that developing policies was a significant or even feasible action of GWIs, in Table 3.13 we find that eight answers (10% of the total) maintained that lasting policies had nonetheless resulted from GWI actions.
- Additionally, although cooperation and collaboration were cited as a significant locus of GWI activity in the previous table (18% of all responses), it accounted for only 4% of the responses when participants were discussing observed results.

- As with Table 3.12, indecision and imprecision, probably a proxy for uncertain expectations and scepticism about the value of GWIs, is striking: 24 of the total 77 responses were of this general type.
- Not surprisingly, amid this uncertainty, only a single respondent explicitly claimed that GWIs had made a "real difference on the ground" by improving quality of life while another stated that "stakeholder needs are rarely directly benefited by international programmes except in a "token" way or through 'demonstration projects'" (W. J. Shuttleworth).
- One respondent suggested a possible explanation for the patterns of non-specificity and pessimism when he claimed that "few if any of the initiatives have actually been tested against measures of success and where this has been attempted the results have not met the stated objectives" (J. Rodda). Another respondent, less diplomatic in his assessment, stated flatly that "I do not think any individual process or initiative is really having a 'major' impact on the global water situation right now" (J. Kuylenstierna).

Table 3.13 Programme results

Categories of responses 33 persons 77 responses	Responses [% (number)]
Advancing ideas and practices	43% (33)
Publications	14 (11)
Projects and programmes	9 (7)
Meetings and conferences	4 (3)
Unspecified	16 (12)
Developing infrastructure	27% (21)
Guidelines, models and toolkits	10 (8)
Research and data	9 (7)
Institutional	6 (5)
Technology	1 (1)
Influencing policy	10% (8)
National	5 (4)
International	3 (2)
Institutional	1 (1)
Unspecified	1 (1)
Cooperation and collaboration	4% (3)
Improved quality of life	1% (1)
No results reported	14% (11)
"Too early to tell"	10 (8)
Unspecified	3 (2)
No significant results	1 (1)

Evaluation of a sample of 41 of the 52 returned long forms (Survey 2). 14 January 2005.

Institutional Overlap: Observations on Table 3.14

Donors claim to be confused, but the situation is no different than in other sectors.

– Margaret Catley-Carlson, Chair, Global Water Partnership

- In considering the important issue of institutional overlap among GWIs, nearly three times as many answers attempted to explain the impact of overlap (52) as those that sought to define its origins and causes (18). Regarding this asymmetry, one respondent suggested that the phenomenon of overlap in the water sector is not particularly remarkable: "donors claim to be confused, but the situation is no different than in other sectors" (M. Catley-Carlson).
- Many more respondents characterized the relative impact (positive vs negative) of overlap (24) than offered an assessment of its significance (only eight).
- Of those who rated the impact of overlap, 58% considered it negative and only 38% found it positive. Overlap was thus clearly viewed as less than desirable.
- Consistent with the above observation, 15 of 20 responses (75%) indicated that overlap was prevalent. (In fact, in response to the invitation to identify specific examples, one respondent sardonically quipping, "Do you want me to write a book?!") Furthermore, since, as was indicated above, overlap is undesirable, we can infer that it was seen by them to be *too* common.
- Half of those who rated the scale of the impact thought it significant.
- Those respondents who offered commentary on the origins of overlap pointed to several sources. For example, respondents suggested that the failure of existing initiatives to cooperate, collaboration and coordinate their activities has led to duplication of programmes when different GWIs work independently to develop essentially the same programmes (instead of, e.g., working together to develop shared institutional infrastructures).
- Another contention (one often heard repeatedly in face-to-face interviews) was that sometimes new GWIs are created for what amounts to personal reasons, namely to fulfil the aspirations or ambitions of forceful individuals, or as one individual wrote, "I think that it is too often a matter of people trying to build empires for themselves (not least in some of the 'new' NGOs)". Table 3.14 confirms that such a factor exists; five responses (28% of those who provided reasons) suggested that one reason for overlap was the draw of power and influence.

Table 3.14 Institutional overlap

Categories of responses 32 persons 70 responses	Responses [% (number)]
Origins of and reasons for overlap	100% (18)
Poor coordination, cooperation, and collaboration	28 (5)
Using water politics as an entrée to power and influence	28 (5)
Financing	16 (3)
Other constraints or deficiencies	22 (4)
"Spin-off" from parent organizations	6 (1)

Table 3.14 (Continued)

Aspects of overlap	
Perception of impact of overlap	100% (24)
Negative	58 (14)
Positive	38 (9)
Neutral	4 (1)
Prevalence of overlap	100% (20)
Common	75 (15)
Rare	20 (4)
Non-existent	5 (1)
Scale of impact	100% (8)
Significant	50 (4)
Minor	38 (3)
None	12 (1)

Evaluation of a sample of 41 of the 52 returned long forms (Survey 2), 14 January 2005.

Proliferation: Observations on Table 3.15

None of these bodies wishes to surrender elements of its own programme for the common good.

– John Rodda

- Table 3.15 shifts the issue from overlap to proliferation. We have already seen that attitudes towards overlap are generally negative. Table 3.15 extends and amplifies this characterization when considering the creation of new GWIs. Of the 46 answers that depicted perceptions of the impact of proliferation, nearly two-thirds of them (64%, as compared with 58% in Table 3.15) interpreted the trend negatively. One respondent explained the danger of proliferation in terms of the fragmentation of a non-proliferating resource pool: "We have mandates by the room full and yet the financial resources with which to address problems remains constant, at best" (R. Meganck).
- Of the 18 responses regarding the scale of the impact, 56% saw it as significant (vs 50% in Table 3.14).
- Expectedly, along the same lines, 9 of the 10 answers that cited intensified competition as one of the effects of proliferation, saw such competition as a drawback, not a benefit. One respondent captured this sentiment when he remarked that "None of these bodies wishes to surrender elements of its own programme for the common good. There is little contact between associations; there is no sense of the need to share and competition is widespread. Despite associations claiming certain scientific territories, others will 'rustle in' and try to take over. The forces are similar to the demographic pressures evident in the world at large" (J. Rodda).
- Further, of the other six types of effects identified, only two ("opportunities for existing initiatives" and "niche specialization") could be considered beneficial;

these drew eight citations (22% of those who identified effects). In other words, 78% of the effects listed were mainly negative.

- To sum up, Table 3.14 indicates that respondents found overlap of GWIs problematic and undesirable. When asked more specifically about proliferation, which can be seen as a sort of intensification of overlap, according to Table 3.15, respondents became even more negative.

Table 3.15 Proliferation

Categories of responses 34 persons 83 responses	Responses [% (number)]
Types of effects	100% (37)
Intensified competition	
Drawbacks (9)	27 (10)
Benefits (1)	
Additional work or activity for GWIs	22 (8)
Confusion, disorder, or added complexity	16 (6)
Opportunities for existing initiatives	14 (5)
Distraction	11 (4)
Niche specialization	8 (3)
Slowing implementation	3 (1)
Relative impacts	
Perception of impact	100% (28)
Negative	64 (18)
Positive	25 (7)
Neutral	11 (3)
Scale of impact	100% (18)
Significant	56 (10)
Minor	28 (5)
None	16 (3)

Evaluation of a sample of 41 of the 52 returned long forms (Survey 2), 14 January 2005.

Managing Diversity and Proliferation: Observations on Table 3.16

Overlap will never be eliminated, particularly when every issue can legitimately be claimed by almost any sector or group. But without GWIs, more chaos would prevail.

– Richard Meganck, Rector, Institute for Water Education (IHE)

Although further fragmentation takes place, strengthening of existing initiatives is important.

– C. D. Thatte, Secretary General ICID

- After definitively opining that overlap is not beneficial and that proliferation is worse, Table 3.16 shows a startling and apparently contradictory view: in choosing between "guiding" proliferation or limiting it, 47 of 57 responses (82%) were to guide it. Put another way, there may be too many GWIs, but by and large, their proliferation should not be limited; instead their efforts should be steered in ways to derive maximum benefit from their actions. One respondent addressed the paradox as follows: "Overlap will never be eliminated, particularly when every issue can legitimately be claimed by almost any sector or group. But without GWIs, more chaos would prevail" (R. Meganck).

- Even of the 10 responses that favoured some limits, about half called for less-than-drastic attempts to decrease the proliferation rate while the remainder suggested merging overlapping organizations and eliminating the inefficient and superfluous. In the latter vein, one respondent wrote that "sunset clauses should be compulsory in all new global water initiatives, the GWIs should go away automatically unless there is a real need for them to continue" (W. J. Shuttleworth).
- In short, as the strategies listed in the top half of Table 3.16 illustrate, respondents perceived that flexible management of organizational overlap is likely the most appropriate option for dealing with proliferation. Or, as another respondent stated, "Although further fragmentation takes place, strengthening of existing initiatives is important" (C. D. Thatte).
- To paraphrase Malin Falkenmark, GWIs mirrors the existence of numerous, perhaps too many, species on our planet. Just as all species benefit from the diversity that results from what may be overproliferation, so institutions such as GWIs may benefit from the richness and variety of approaches, opinions and individuals.

Table 3.16 Managing diversity and proliferation

Categories of responses 28 persons 57 responses	Responses [% (number)]
Guiding diversity	82% (47)
Maximize existing opportunities for collaboration	25 (14)
Create or designate overarching authority	12 (7)
Leverage resources to promote cooperation	7 (4)
Continue existing agendas and approaches	7 (4)
Involve politicians and potential adversaries	7 (4)
Facilitate information-knowledge exchange	5 (3)
Develop new approach or framework	5 (3)
Decrease collaboration "transaction costs"	4 (2)
Clarify group roles and boundaries	4 (2)
Non-specific	7 (4)
Limiting proliferation	18% (10)
Decrease rate of GWI proliferation	9 (5)
Eliminate inefficient and superfluous GWIs	5 (3)
Merge overlapping GWIs	4 (2)

Evaluation of a sample of 41 of the 52 returned long forms (Survey 2), 14 January 2005.

Observations on Table 3.17: Assessing Overall Impact

Certainly GWIs have had very substantial results in terms of scientific understanding and some socioeconomic benefits; the transition of this understanding into practical benefit in individual catchments has been less successful.

– W. J. Shuttleworth, Professor, University of Arizona

At least these institutions provide some context to reorient the decision processes to support the importance of water in economic development.

– Richard Meganck

- When asked to assess overall impact, that is aggregate influence and success, and in spite of their concerns about overlap and proliferation, 23 of 29, or 48%, of respondents expressed the opinion that GWIs had exerted positive or at least partially positive influence.
- Of those who stated which kinds of influences were most significant, a quarter discussed "real" changes; but of those fewer than half thought GWI efforts had been successful. This sense of ambivalence was captured by one respondent who opined that while "Certainly GWIs have had very substantial results in terms of scientific understanding and some socio-economic benefits; the transition of this understanding into practical benefit in individual catchments has been less successful" (W. J. Shuttleworth).
- When turning from concrete, measurable changes to fuzzier ones such as increased awareness, one-fifth of all responses indicated that GWIs' efforts to enhance awareness had been successful, with one respondent writing that "at least these institutions provide some context to reorient the decision processes to support the importance of water in economic development" (R. Meganck).
- Results were very similar with regard to improving communication and cooperation. There, six of seven agreed that such work had succeeded.
- With respect to facilitating the work of those involved in the water sector, another difficult-to-measure achievement, seven of eight answers affirmed that this was an outcome of GWI activity.
- Finally, of the six respondents who commented on the impact of GWIs on the water sector's orderliness and efficiency, half suggested that GWIs have failed to significantly improve the sector's efficiency.
- Overall, the responses indicate that GWIs are perceived to have accomplished a series of intermediate goals but have not yet been successful in leveraging these to achieve their ultimate, concrete goals. As one thoughtful respondent preffered: "Actual benefits derived from the work of GWIs are subtle and elusive: support for ongoing projects, sharpening and disseminating the rhetoric of international water agendas, and sensitizing national and sub-national governments to (1) long-term threats, (2) available instruments to resolve issues, (3) evolution of international water law, (4) availability of technical assistance. Yet, these important developments may result in concrete successes only after many years have passed..." (M. Reuss).

Table 3.17 Assessing overall impact

Categories of responses 29 persons 72 responses	Responses [% (number)]
GWIs exerted positive influence	100% (29)
Yes	48 (14)
To an extent, or partially	31 (9)
Not really	21 (6)
Kinds of influence	100% (38)
Produces "real" or concrete change	24 (9)
Successful	(4)
Unsuccessful	(5)
Increased awareness or status of water issues	21 (8)
Successful	(8)
Unsuccessful	(0)
Facilitated work of actors in water sector	21 (8)
Successful	(7)
Unsuccessful	(1)
Increased communication, cooperation and collaboration	18 (7)
Successful	(6)
Unsuccessful	(1)
Increased order or efficiency in water sector	16 (6)
Successful	(3)
Unsuccessful	(3)
Caveats	100% (5)
Long-term (not short-term) benefit	40 (2)
Effectiveness is scale and context dependent	40 (2)
Need to evaluate opportunity costs of GWIs	20 (1)

Evaluation of a sample of 41 of the 52 returned long forms (Survey 2), 14 January 2005.

Lessons Learned: Observations on Table 3.18

Urgent problems demand quick responses, and global water initiatives are usually not the vehicle of choice to resolve these problems.

– Martin Reuss, Historian, United States Army Corps of Engineers

The instruments available to nation-states today are not adequate for dealing with global and national institutional problems relating to water, natural resources and the environment.

– John Rodda

There are too many overlaps and poor coordination between the plethora of initiatives, which so far have yielded little positive change.

– Peter Bridgewater, Secretary General, Ramsar Convention on Wetlands

There is a huge disconnect between the global discourse and real-world water management on the ground there is very little impact in the real world.

– Aaron Wolf

- The large number of responses (78) suggests that respondents were eager to draw lessons from the experiences of GWIs.
- If the responses can be summarized as a coherent recommendation, it would be that the key to GWI success would be to streamline organizational practice so as to facilitate innovation and, especially, cooperation and collaboration with other organizations and with relevant stakeholders.
- The importance of enhancing cooperation and collaboration was cited most often by respondents, with 36% suggesting various ways to achieve this. Regarding this need, one respondent offered the following observation: "There are too many overlaps and poor coordination between the plethora of initiatives, which so far have yielded little positive change" (P. Bridgewater).
- Meanwhile, the goal of encouraging innovation accounted for 19% of all responses, with one respondent commenting: "The instruments available to nation-states today are not adequate for dealing with global and national institutional problems relating to water, natural resources and the environment" (J. Rodda).
- The need to streamline organizational practice was expressed most frequently (29% of all responses) in terms of having to improve the precision and specificity of practice through the standardization of techniques and procedures, addressing problems at the appropriate temporal and geographic scale, and adopting appropriate and specific roles. For example, regarding the issue of temporal scale and GWIs' role, "urgent problems demand quick responses, and global water initiatives are usually not the vehicle of choice to resolve these problems. National and sub-national states generally do *not* depend on global water initiatives to resolve immediate problems, although they may use data obtained in earlier global efforts to enhance their arguments" (M. Reuss).
- A smaller number of responses (13% of the total) discussed streamlining of organizational practice through their identification of a number of important institutional tensions or disjuncture requiring redress in order to improve GWI performance. One respondent noted, for example, a disjuncture between rhetoric and practice when he wrote, "there is a *huge* disconnect between the global discourse and real-world water management on the ground; there is *very* little impact in the real world" (A. Wolf).
- As all four of the major divisions of Table 3.18 suggest, once again it has been the "softer", less-easily quantifiable aspects of GWI operation that have drawn the most attention from respondents.

Table 3.18 Lessons learned

Categories of responses 31 persons 78 persons	Responses [% (number)]
Enhancing cooperation or collaboration	36% (28)
Networking	21 (16)
Promoting participation	9 (7)

Table 3.18 (Continued)

Sustaining communication among members	4 (3)
Promoting transparency and trust	3 (2)
Improving precision and specificity	29% (23)
Standardization of methods, formats, protocols	13 (10)
Use of appropriate temporal and geographic scales	12 (9)
Better-defined organizational roles and niches	5 (4)
Encouraging innovation	19% (15)
Organizational and institutional	8 (6)
Alignment of concepts and practices	5 (4)
Technological	4 (3)
Significance of understanding of history	3 (2)
Addressing institutional tensions	13% (10)
Models vs "real world"	4 (3)
Rhetoric vs practice	4 (3)
Research vs management	3 (2)
Costs vs Benefits of Networks	1 (1)
Capital vs Ideas as Principal Policy Driver	1 (1)
"Too soon to tell"	3% (2)

Evaluation of a sample of 41 of the 52 returned long forms (Survey 2), 14 January 2005.

3.6 Conclusions

Following a detailed discussion of the character and history of a particular set of global institutions called global water initiatives (GWIs), the authors offered a detailed analysis of a two-part survey administered to leading figures in the field of global water management and research in an effort to elicit their informed perceptions regarding (a) the extent to which each of 30 prominent global water initiatives has influenced the wider "world of water" and (b) roles of different constituencies in GWIs, impacts of GWIs' actions, extents and consequences of overlap and proliferation, overall significances of GWIs for global water management and research, and lessons learned from GWIs.

The four tabulations summarizing Survey 1 and the eight reporting on Survey 2 reveal a number of important insights that would not have been apparent without this analysis. The first group of tabulations enables an enumeration of the many global water initiatives competing for resources, influence and opportunity. In ensuing discussion, the chapter shows that, though diverse, these institutions are bound by a very loose commonality, a fundamental purpose "to advance the knowledge base regarding the world's inland water and its management" (see Introduction). Beyond this, the authors' research suggests the utility of distinguishing between four separate categories of global water initiatives: professional societies, designated periods, organized events and organizations.

Figure 3.2, Tables 3.7, 3.8, 3.9, and 3.10 and associated discussions go further by helping to identify (a) general trends among the wider field of GWIs and (b) which GWIs within each of the four categories were most highly regarded and which seemed to be perceived as having less utility. The data suggest that organized events, as a category, were perceived by respondents as being the most influential while organizations received the lowest rating (mean scores were 2.84 and 3.19, respectively). The average ratings for individual GWIs occupied a range of 2.17 to 3.88.[32] The relatively narrow ranges for both GWIs as categories and individually (0.35 and 1.71, respectively) would suggest that the field of GWIs is one in which no individual element or single class of elements wields significant disproportionate influence. Furthermore, overall, respondents gave GWIs a decidedly mediocre rating (3.03 was the mean score for the entire sample), indicating that, on average, individual initiatives were perceived to have had only moderate levels of influence on the wider "world of water". This would suggest that what influence these initiatives have had has been a product of their *aggregate* force (see below).

Moving from an analysis of the first to the second survey instrument, Tables 3.11, 3.12, 3.13, 3.14, 3.15, 3.16, 3.17, and 3.18 and accompanying observations seek to elicit and assess more nuanced accounts of respondents' perceptions of global water initiatives using a detailed instrument based on an open-ended response format. This analysis provided several key insights.

- First, in Table 3.11 respondents indicated that they perceived governments, NGOs and general stakeholders all to have a place in guiding the work of GWIs, though the roles of each were differently constituted. In particular, governments were valued for providing leadership and funding, while NGOs were seen principally as serving an advisory function and as a mechanism for facilitating broader participation. General stakeholders shared this participatory role and for articulating the agendas of more local constituencies.
- Table 3.12 shows what leaders in the field consider to have been the significant actions taken by GWIs in the past. Half of the responses described efforts to advance more sophisticated ideas and better practices while a third spoke of the generation of institutional, legal, technical and conceptual infrastructures; whereas the remaining one-fifth of responses were dedicated to efforts to promote collaboration among groups within the water sector.
- Table 3.13 shows what respondents perceived to be the results of these efforts. Though roughly symmetrical to the data on actions taken, several anomalies were noted: few noted increased collaboration and cooperation among groups as an outcome while policy change was more often cited as an outcome than as a focus of GWI efforts. Still, a remarkable 20% of responses were non-specific, while 10% indicated that it was too early to discern effects of GWI activity.
- Table 3.14 offers an intriguing snapshot into perceptions of the issue of institutional overlap, suggesting that it is both undesirable and prevalent. Of the relatively few responses offered regarding the origins of this overlap, 28% claimed

[32] The International Conference on Water and the Environment and the GIWA, respectively.

poor coordination, cooperation and collaboration as significant factors while another 28% of responses involved account of how GWIs had served as vehicles for the ambitions of various interested parties.

- Table 3.15 summarizes respondents' perceptions regarding the consequences of the proliferation of initiatives. Responses indicated that the impacts of proliferation were perceived as both negative and of significant scale. Of the specific consequences cited, 75% were negative (including intensification of competition between GWIs, strain on existing initiatives, and escalating confusion, complexity and disorder), while only 25% of responses described positive effects of proliferation (including the creation of new opportunities and trends towards greater specialization).
- Table 3.16 shows how respondents thought these issues might best be dealt with. Significantly, the analysis found that only 18% of responses were constituted by calls to limit proliferation while the remaining 82% offered suggestions on how diversity might be productively guided.
- Table 3.17 summarizes perceptions of the overall impact of GWIs. Results indicate that, in general, GWIs are perceived to have exerted a positive influence (with 79% of responses suggesting that at least some benefit has accrued on account of their existence). In discussing how, specifically, influence has been exerted by GWIs, respondents suggested that they have been successful in raising awareness of water as an issue, increasing cooperation and collaboration, and facilitating the work of water management. However, evaluation of GWIs' relative success in increasing the water sector's efficiency and producing "real," concrete changes on the ground were ambivalent, with responses split roughly 50/50 regarding whether or not they have affected significant positive change.
- Finally, Table 3.18 reveals respondents' perceptions regarding what key lessons have been learned through their experience with GWIs. Responses suggest the importance of streamlining organizational practice, encouraging continued innovation and facilitating cooperation and collaboration as a means of rendering the field of GWIs more efficient and productive.

Having presented and discussed these findings, the authors can now address the chapter's original hypothesis: that the numerous existing GWIs often have duplicative aims and have over-proliferated, and that, therefore, knowledgeable observers would tend to minimize the salutary influences of GWIs and perhaps advocate their consolidation or selective elimination.

As the analysis of Tables 3.14 and 3.15 demonstrates, the first part of the hypothesis is easily confirmed. A clear majority of respondents felt that there is too much overlap among GWIs and that these institutions are multiplying too frequently. Yet, surprisingly, the next two tabulations (Tables 3.16 and 3.17) reveal an unexpected turn: these same respondents did not suggest eliminating or merging competing initiatives. On the contrary, they seemed to accept their existence, embrace their diversity and willingly suggest ways for them to contribute more effectively.

With global water initiatives, it seems, as with grass species or invertebrates, diversity promotes competition and by and large is seen as a healthy attribute. The alternative, as Malin Falkenmark has sardonically suggested, is a monocultural, monolithic model that while perhaps more efficient would surely be less interesting and less progressive (Falkenmark 2004).

Acknowledgements

The present work could not have been accomplished without the benefit of a year's sabbatical granted to Robert Varady by the University of Arizona (UA). There, special thanks go to Stephen Cornell, director of the Udall Center for Studies in Public Policy and Richard Powell, former Vice President for Research and Graduate Studies, for their support, and to W. James Shuttleworth, UA professor of Hydrology and Water Resources, for his encouragement. In addition, the research has been partly funded by a fellowship to Varady from the University's Institute for the Study of Planet Earth.

Our Udall Center colleague Robert Merideth has generously reviewed and edited the manuscript and contributed greatly to its composition. The authors also gratefully acknowledge András Szöllösi-Nagy and Michael Bonell of UNESCO's International Hydrological Programme (IHP). Many others at IHP and elsewhere at UNESCO have been helpful. The late Michel Batisse of UNESCO merits special recognition for his assistance and encouragement. Author Robert Varady also recognizes the influence on his thinking of the HELP (Hydrology for Environment, Life and Policy) Initiative. It is that programme which has been headquartered at IHP and coordinated by Michael Bonell that first drew his attention to the global water initiatives phenomenon.

Additionally, the authors are deeply indebted to more than 70 respondents of Robert Varady's survey. They gave generously of their time, some in personal interviews, and offered invaluable insights into the "world of water".

Finally, we recognize that completion of the present chapter was prompted by the January 2005 workshop in Bangkok that has served as the organizing principle for the present book. In that regard, we thank the Sasakawa Peace Foundation USA and Japan, and the Third World Centre for Water Management for their travel support.

References

Batisse M (2003) Personal Interviews with Robert Varady, Paris, France (October 3, November 28)

Batisse M (2005) Du désert jusqu'B l'eau … 1948–1974; La question de l'eau et l'Unesco: de la 'Zone aride' B la 'Décennie hydrologique.' Paris: Club Histoire; Association des anciens fonctionnaires de l'Unesco

Bernard R (1995) Research Methods in Anthropology, 2nd Edition. Altamira Press, NY

Biswas AK (2002) Personal Communication with Robert Varady (September 7)

Brewster M (2004) International Year of Freshwater: the Achievements and Successes of Similar International Initiatives. In: Rodda JC, Ubertini L (eds) The Basis of Civilization—Water Science? IAHS Press, Oxfordshire, UK

Calder J (1996) Statistical Techniques. In: Sapsford R, Jupp V (eds) Data Collection and Analysis. Sage, London, pp. 226–261

Chapman S (1959) IGY: Year of Discovery. University of Michigan Press, Ann Arbor

Cosgrove WJ (1999) A Vision for Water, Life and the Environment in the 21st Century. Keynote Address, UNED-UK Conference: Setting the Freshwater Agenda for the 21st Century (16 June)

Cosgrove WJ, Rijsberman FR (2000) World Water Vision: Making Water Everybody's Business. Earthscan Publishing, London

Day JBW (1999) A Brief History of IAH. Unpublished Adaptation of Account Published in Applied Hydrologeology 1, 1 (1992) 2pp.

Delli Priscoli J (1998) Water and Civilization: Using History to Reframe Water Policy Debates and to Build a New Ecological Realism. Water Policy 1: 623–636

Delli Priscoli J (2005) Personal communication (April 8)

De Vellis FF (1991) Scale Development: Theory and Applications. Sage, Newbury Park, CA

The Dublin Statement on Water and Sustainable Development (1992). www.wfeo-comtech.org/WorldWaterVision/DublinStatementH20AndSD.html (September 22, 2005)

Entekhabi D, Asrar GR, Betts AK, Beven KJ, Bras RL, Duffy CJ, Dunne T, Koster RD, Lettenmaier DP, McLaughlin DB, Shuttleworth WJ, van Genuchten MT, Wei MY, and Wood EF (1999) An Agenda for Land Surface Hydrology Research and a Call for the Second International Hydrological Decade. Bulletin of the American Meteorological Society 80(10): 2043–2058

Falkenmark M (2001) Ten Year Message from Previous Symposia. Water Science and Technology 43(4): 13–15

Falkenmark M (2004) Personal Interview with Robert Varady (May 3). Stockholm, Sweden

George CB (2003) Personal Communication Between Author and Executive Director of IAHR (October 3)

GIWA (2005) International Waters Assessment. www.giwa.net/index.phtml (September 22)

Healy JF (1990) Statistics: A Tool for Social Research. Chapman and Hall, London

HELP (2005). (Hydrology for Environment, Life and Policy) Portal.unesco.org/sc_nat/ev.php?URLID=1205&URL_DO=DO_TOPIC&URLSECTION=201 (September 22)

HELP (2001) (Hydrology for Environment, Life and Policy) Task Force. The Design and Implementation Strategy of the HELP Initiative. UNESCO, Paris

Keohane RO, Haas PM, Levy MA (1994) The Effectiveness of International Environmental Institutions. In: Haas PM, Keohane RO, Levy MA (eds) Institutions for the Earth: Sources of Effective International Environmental Protection. The MIT Press, Cambridge, MA, pp. 3–24

Korzoun VI (1991) 25 Years of IHD/IHP. In: International Symposium to Commemorate the 25 Years of IHD/IHP. Convened by UNESCO 15–17 March 1990. UNESCO, Paris, pp. 13–18

Korzoun VI (ed) (1978) World Water Balance and Water Resources of the Earth. Studies and Reports in Hydrology. UNESCO, Paris, 663pp.

Kuylenstierna JL (2003) The World Water Forum in Kyoto – Just Another Conference. Stockholm Water Front: A Forum for Global Water Issues 1 (April): pp. 10–11

Likert R (1932) A Technique for the Measurement of Attitudes. Archives of Psychology 140: 5–53

Marsh C (1988) Exploring Data: An Introduction to Data Analysis for Social Scientists. Polity Press, Cambridge

Najlis P, Kuylenstierna JL (1997) Twenty Years After Mar Del Plata—Where Do We Stand and Where Do We Go? In: Proceedings, Mar Del Plata 20 Year Anniversary Seminar: Water for the Next 30 Years—Averting the Looming Water Crisis, pp. 23–35. Stockholm, Sweden (August 16) Stockholm International Water Institute, Stockholm

Rodda JC (1991) Speech on behalf of the United Nations System. In: International Symposium to Commemorate the 25 Years of IHD/IHP (March 15–17, 1990). UNESCO, pp. 5–8. UNESCO, Paris

Rodda JC (1999) Promoting International Co-operation in Learning—The Role of the International Association of Hydrological Sciences. In: Proceedings of the International Symposium, The Learning Society and the Water-Environment. Ed. By A. Van der Beken, M. Mihailescu, P. Hubert, and J. Bogardi. Paris, France, 2–4 June 1999

Rodda JC, Ubertini L (eds) (2004) The Basis of Civilization—Water Science? IAHS Press, Oxfordshire, UK

Salman-Salman MA (2003) From Marrakech Through The Hague to Kyoto: Has the Global Debate on Water Reached a Dead End? Part One. Water International 28(4): 491–500

Salman-Salman MA (2004) From Marrakech Through The Hague to Kyoto: Has the Global Debate on Water Reached a Dead End? Part Two. Water International 29(1): 11–19

Spector PE (1992) Summated Rating Scale Construction. Sage, Newbury Park, CA

Speth JG (2003) Perspectives on the Johannesburg Summit. Environment 45(1): 25–29

Udall SN, Varady RG (1993) Environmental Conflict and the World's New International Borders. Transboundary Resources Report 7(3):5–6

UNACC, ISGWR (United Nations Administrative Committee on Coordination) (Inter-Secretariat Group for Water Resources) (1992) The Dublin Statement and Report of the Conference. International Conference on Water and the Environment: Development Issues for the 21st Century, 55pp. (January 26–31). Dublin, Ireland

UNESCO, WMO (World Meteorological Organization) (1988) Third UNESCO/WMO International Conference on Hydrology and Scientific Bases of Water Resources Management, 232pp. Paris, France, Geneva, Switzerland

Varady RG (2003) Global Water Initiatives: Some Preliminary Observations on Their Evolution and Significance. Presented at 3rd Conf. of International Water History Association 25pp, Alexandria, Egypt, December 11–13

Varady RG (2004) Global Water Initiatives: Some Observations on Their Evolution and Significance. Proceedings of the AWRA & IWLRI International Specialty Conference on Good Water Governance for People and Nature. What Roles for Law, Institutions, Science and Finance? 9pp.

Victor DG, Skolnikoff EB (1999) Translating Intent into Action. Environment 41(2): 16–20, 39–43

Volker A, Colenbrander HJ (1995) History of the International Association of Hydrological Sciences. IUGG Chronicle 94(1): 13–22

Weiner D (1992) Demythologizing Environmentalism. Journal of the History of Biology 25(3): 385–411

Wolf AT (1998) Conflict and Cooperation along International Waterways. Water Policy 1: 251–265
WWAP (World Water Assessment Programme) (2003a) The United Nations World Water Development Report. UNESCO & Berghahn Books, Paris, 576pp.
WWAP (World Water Assessment Programme) (2003b) Milestones. In: The United Nations World Water Development Report, pp. 24–28
WWAP (World Water Assessment Programme) (2005). www.unesco.org/water/wwap/index.shtml (September 22)

Appendices

Appendix A

Survey 1: Influences and Connections (for "Observers")

Please indicate with an "X" in the appropriate box (one value only)

1 = Very strong . . . 3 = Moderate . . . 5 = Very weak to non-existent

1. Events

Rank the intensity of the influence of these events and their outcomes:	1	2	3	4	5
1965–1974 International Hydrological Decade (IHD)					
1972 UN Conference on the Human Environment (Stockholm)					
1977 UN Conference on Water (Mar del Plata)					
1981–1990 International Drinking Water Supply & Sanitation Decade					
1992 International Conference on Water and the Environment (Dublin)					
1992 UN Conference on Environment and Development (Rio Earth Summit)					
1997 First World Water Forum (Marrakech)					
1998 International Conference on Water and Sustainable Development (Paris)					
2000 Second World Water Forum (the Hague)					
2000 Millennium Development Goals					
2001 International Conference on Freshwater (Bonn)					
2003 International Year of Freshwater					
2003 Third World Water Forum (Kyoto)					
Other event					
Other event					

2. Other Institutions

Rank the intensity of the influence of each of the following institutions:	1	2	3	4	5
Dialogue on Water, Food and Environment					
Global Energy and Water Cycle Experiment (GEWEX)					
Global International Waters Assessment (GIWA)					
Global Water Partnership (GWP)					
Hydrology for Environment, Life and Policy (HELP) Initiative					
Stockholm International Water Institute (SIWI)					
UNESCO's International Hydrological Programme					
Water Supply and Sanitation Collaborative Council (WSSCC)					
World Water Assessment Programme (WWAP)					
World Water Council (WWC)					
Global Water System Project (GWSP)					
Other institution					
Other institution					

3. Professional Associations

Rank the influence of each of the following professional associations:	1	2	3	4	5
IAHR (International Association for Hydraulic Research)					
International Association of Hydrogeologists (IAH)					
International Association of Hydrological Sciences (IAHS)					
International Water Association (IWA)					
International Water Associations Liaison Committee (IWALC)					
International Water Resources Association (IWRA)					
Other professional association					

Survey 1: Influences and Connections (for "Representatives")

1 = Very strong . . . 3 = Moderate . . . 5 = Very weak to non-existent

1. Events

Rank the intensity of the influence on your institution of these events and their outcomes:	1	2	3	4	5
1965–1974 International Hydrological Decade (IHD)					
1972 UN Conference on the Human Environment (Stockholm)					
1977 UN Conference on Water (Mar del Plata)					
1981–1990 International Drinking Water Supply & Sanitation Decade					
1992 International Conference on Water and the Environment (Dublin)					
1992 UN Conference on Environment and Development (Rio Earth Summit)					
1997 First World Water Forum (Marrakech)					
1998 International Conference on Water and Sustainable Development (Paris)					
2000 Second World Water Forum (the Hague)					
2000 Millennium Development Goals					
2001 International Conference on Freshwater (Bonn)					

2003 International Year of Freshwater
2003 Third World Water Forum (Kyoto)
Other event
Other event

2. Other Institutions

Rank the intensity of the influence on your institution of each of the following institutions:	1	2	3	4	5
Dialogue on Water, Food and Environment					
Global Energy and Water Cycle Experiment (GEWEX)					
Global International Waters Assessment (GIWA)					
Global Water Partnership (GWP)					
Hydrology for Environment, Life and Policy (HELP) Initiative					
Stockholm International Water Institute (SIWI)					
UNESCO's International Hydrological Programme (IHP)					
Water Supply and Sanitation Collaborative Council (WSSCC)					
World Water Assessment Programme (WWAP)					
World Water Council (WWC)					
Global Water System Project (GWSP)					
Other institution					
Other institution					

3. Professional Associations

Rank the intensity of the connection of your institution to each of the following professional associations:	1	2	3	4	5
IAHR (International Association for Hydraulic Research)					
International Association of Hydrogeologists (IAH)					
International Association of Hydrological Sciences (IAHS)					
International Water Association (IWA)					
International Water Associations Liaison Committee (IWALC)					
International Water Resources Association (IWRA)					
Other professional association					

Appendix B

Survey 2: Global Water Initiatives (for "Observers")

Over the past decades, the global water agenda has been shaped by numerous institutions. Among these, *professional societies* such as the International Association of Hydrological Sciences (IAHS) and the International Association of Hydrogeologists (IAH), *dedicated time periods* such as the International Hydrological Decade, *organizations* such as the World Water Council (WWC) and the Global Water Partnership (GWP), and *specialized events* such as the 1977 UN Conference on Water and the more recent World Water Forums all have actively worked to influence research, development and implementation of water-related programmes.

As someone who has observed and participated in events related to the above phenomenon, would you respond to the following contextual and analytical questions?

1. Intellectual currents

- In your opinion, what intellectual currents and big ideas have prompted and shaped the evolution of global water initiatives?
- In what ways are these products of the thinking of the time?
- Who have been the leaders and visionaries?
- How have the ideas evolved?

2. Organizational background

- Which of the types of institutions mentioned earlier are you most familiar with? Please identify specific ones and insofar as possible, refer to those in your subsequent responses.
- What are their principal orientations (e.g., disciplinary, sectoral)?
- What are their overall goals or visions?
- What have been their main driving forces?
- How have they been financed? At what levels?
- How have these initiatives evolved their organizational strategies, intellectual orientations or practical foci?
- What, if any, has been *your* personal role?

3. Practical currents

- What do you think is the approach of these initiatives to multilateral resource-governance?
- What have been the sociopolitical drivers? How do these reflect global, regional and national politics?
- To what degree are products and results emphasized?
- In what ways or circumstances are global water initiatives subject to diplomatic or political constraints?

4. Governance

- How have the above influences been manifested in practical institutional design?
- What do you think are the most successful structural/organizational models?
- What is the role of governments in the work of global water initiatives? NGOs? Stakeholders?
- What "communities" have been formed among participants and others, such as other decision-makers, academics and activists?

5. Evaluation of success/failure

- What have been the most important actions taken by the initiatives you are most familiar with? What prompted these?
- Who are the expected beneficiaries?
- By and large, do you believe that the resources expended have been influential in shaping water-policy agendas?
- Can you name real policy changes that can be attributed to the programmes you know best?
- Do you believe these programmes have yielded lasting, positive and practical results? Can you give examples?
- How have progress and achievements been assessed or measured?
- What have been the perceptions of global water initiatives among governments, managers, stakeholders, academics and other potential beneficiaries? How have these changed?
- What have been some key lessons learned by water initiatives?
- Are there procedures to make the institutional designs responsive to these lessons?

6. Institutional overlap, cooperation and competition

- How are existing initiatives affected by the continual creation of new initiatives?
- Which of these do you see as most influential in affecting the global water situation?
- Can you identify overlaps, opportunities for cooperation, and examples of conflicts among these institutions?
- How might the broad policy and global-water-initiatives process be steered to harness similar but not identical agendas among institutions?

7. Big picture/summary

- Are the instruments available to nation-states today adequate for handling global institutional problems related to the environment and natural resources?
- In the case of water, are individual countries able to transcend national interests without the involvement of global water initiatives?
- To the extent that you are familiar with such initiatives, how would you gauge their influence and success? Put another way, how would the "world of water" be different today if global water initiatives did not exist?

8. Other information

- Are there other questions I should be asking?
- Can you suggest written sources?
- Who else should I talk to?

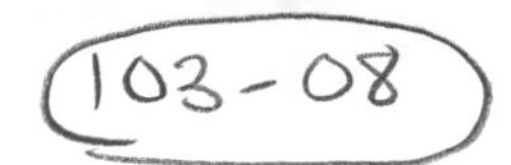

4 Global Water Conferences: A Personal Reflection

Jon Lane

4.1 Introduction

This analysis contains personal reflections on some global water conferences that took place from 1997 to 2003. This period witnessed an unusually large number of such conferences that, taken together, not only made a lot of progress on establishing global aims and priorities for the water sector but also received a lot of criticism for their high cost and low impact. My own background is in the NGO sector, working in drinking water and sanitation in developing countries, and I was fortunate enough to attend several of these global conferences. I have not attempted to write an objective evaluation of them all, but a subjective narrative.

4.2 First World Water Forum, Marrakech, 1997

This was not the first big global conference on water (that accolade belongs to the UN Conference on Water at Mar del Plata back in 1977) but most people recognize it as the first of the recent series of such conferences. It was the World Water Council's first conference, and the organizers clearly decided to play it safe by designing the agenda without any question or discussion time whatsoever. We all sat through three solid days of speeches in plenary, only to be startled from our slumbers at the very end by a stirring speech from the podium beginning "We, the people gathered here…".

Had we discussed or agreed it? No, we had not. But that seemed to make no difference. The speaker (himself a top WWC office-bearer) applauded WWC for hosting such a successful conference, invited it to host a series of them in the future, and requested it to commission a world water vision in time for the next Forum 3 years later. WWC graciously agreed to all these requests, we levered ourselves out of our armchairs and the Forum was triumphantly completed. The wheels were set in motion – and 8 years later they are still inexorably turning.

4.3 International Conference on Water and Sustainable Development, Paris, 1998

I doubt the purpose of this conference was clear to us, participants at the time, and the passing years have made it no clearer to me. The turnout was impressive and the lunches sumptuous, but it seemed to have no connection to any processes or conferences either before or after it.

4.4 Second World Water Forum, The Hague, 2000

Thanks to the energy and commitment of the World Water Council and the Dutch Government, over 5000 participants attended the Forum. This was a far bigger group than at any previous water sector conference. All the regular international conference-goers were there, plus a large number of people, coming particularly from NGOs and civil society, who do not normally attend these global events. Plenty of politicians were also there, attending a Ministerial Conference in the same building at the same time. I was excited to see such a wide range of people gathered and ready to listen, debate and learn. Plenary sessions were sensibly kept to a minimum and for most of the week there was a complex programme of events to interest and involve everybody. However, the sheer size of the Forum brought its own problems – the vast choice of simultaneous activities divided up the group and reduced most of us to attending the sessions on our own subjects. Consequently everybody could have come away feeling their point of view had prevailed, but without even meeting the people who opposed it.

So what were all these people talking about? On this, I sensed some confused planning. On the one hand, the Dutch Government had ensured that every conceivable water-related subject was aired and all viewpoints welcomed. On the other hand, the co-host World Water Council's main plenary sessions and the official publications favoured a right-wing agenda of privatization, economic valuation of water and the power of the global market.

The conflict between this ideology and the left-wing agenda of human rights, pluralism and democratic accountability is a crucial global debate that extends far beyond the confines of the water sector. It underlay many of the discussions at the Hague. Some of the discussions were unnecessarily antagonistic on both sides, but the arguments themselves were important and significant. By the end of the week, I sensed that the globalization advocates did not wield such an influence over the politics of water as they may have wished. In both the Forum and the Ministerial Conference, the advocates of human development made some powerful points.

The main debating points were on the role of the private sector, on the politics of dams, and on balancing the economic and the social value of water. I know that human rights and targets were delicately avoided and that the Ministerial Declaration says little of substance if you read it carefully. I know we were swamped by papers and spoilt by sheer choice of sessions. But many people and organizations received some real insights and some real jolts to their work, the debates were opened and broadened, the media covered water issues as never before, and many practitioners were able to tell others about their work, to influence and to learn.

The World Water Council, as organizer of the Forum, was criticized for its aloofness and its lack of transparency. It had certainly made a muddle of inviting so many different organizations to produce overlapping Visions and Frameworks for Action, which were extremely confusing to the participants at the Hague. The Global Water Partnership was also criticized, but to a lesser degree, and it was recognized as being the central player in the important future debates that will enable the right-wingers and left-wingers, as I have described them earlier, to work together. It took the criticisms to heart, and was making a genuine attempt to improve its openness and its partnerships with others. Of the other global organizations, the Water Supply and Sanitation Collaborative Council was praised for its consultative and participatory processes.

4.5 Millennium Summit, New York, 2000

I did not attend this conference, as it was a formal part of the UN intergovernmental process, but include it here because it adopted the Millennium Development Goals, including the well-known targets for water supply and water resources management (though, disappointingly at the time, not for sanitation).

4.6 International Conference on Freshwater, Bonn, 2001

There had not been an official UN conference dedicated to water since Mar del Plata in 1977. There had, of course, been plenty of non-UN water conferences. The German government therefore designed the Bonn Conference as a one-off event with a particular purpose, namely to build a bridge between the non-UN water meetings and the UN process leading to the Johannesburg Summit the following year. Thus Bonn was a conference about other conferences rather than about ideas. Indeed, coming only 18 months after the Hague Forum, it would have been surprising to find either new subjects to discuss, or much progress on existing subjects.

Some 2000 people from diverse backgrounds attended the Bonn conference: national delegations from both developing and industrialized countries, UN and other multilaterals, NGOs from all parts of the world, private sector, academics, the media. Its hybrid nature felt odd: many of the national delegations comprised legal and diplomatic staff whose jobs seem to consist largely of arguing over wording. Alongside them, the water professionals who normally attend conferences to discuss the actual subject felt slightly out of place.

The subject of dams, which had provoked lively debate and colourful demonstrations at the Hague, scarcely featured in Bonn. In the meantime the World

Commission on Dams had produced its report, and it seemed that the various factions, having failed to make the leap of faith to accommodate each other's views, had retreated to their previous positions.

More positively, there was a genuine determination to regard water as a force for reconciliation and cooperation, as proposed by Kader Asmal in his powerful address at the 2000 Stockholm Water Symposium, rather than as a source of conflict, as balefully foretold by Ismail Serageldin and others in previous years. Gender, too, was well addressed. The problems of corruption, which are known to hang over many water-related transactions around the world, were more overtly discussed that at any previous such conference.

The Bonn Ministerial Meeting was intended to give leadership and direction to the rest of the conference, but it produced a bland statement, ground out by the civil servants in an all-night session. There were signs of dissatisfaction with it from some ministers personally pressing for specific targets for sanitation. The conference itself produced a set of recommendations, as an indication of the consensus feelings, to submit to the preparatory process for Johannesburg. So in that respect it achieved its aim of bringing the results of all our previous deliberations into the UN System.

4.7 World Summit on Sustainable Development, Johannesburg, 2002

This blockbuster event had it all: a formal summit meeting of heads of state cocooned in a conference centre packed with delegates talking non-stop into their mobile phones, vast crowds of media and civil society activists, an entire cultural village at a cricket stadium, wonderful and exuberant South African hospitality, and a large if distant venue devoted entirely to water.

The Water Dome alone was bigger than many other global conferences, but produced mixed reactions in the visitors. It was so far from the summit venue that very few politicians or their advisors visited it, so we mostly talked among ourselves, and we didn't really have anything new to say. At the summit itself, the global media generated grumpily negative reports but we were delighted by one of the few progressive outcomes, namely the adoption of the sanitation target. The US delegation's opposition to it generated much more publicity for the subject of sanitation than if it had simply been nodded through in the agreed text.

4.8 Third World Water Forum, Kyoto, 2003

Normally each meeting in a series builds on the momentum from the previous one. In the case of the Third World Water Forum in Kyoto, this was always going to be a problem, and not just because the previous Forum at the Hague had been such a dynamic and progressive event. The organizers' difficulty arose from the sheer number of global water meetings that had taken place during the intervening 3 years.

After all those global meetings, the Kyoto Forum was always likely to be an anticlimax, at least in respect of policy progress. So its organizers billed it as a showcase for action. It is a difficult task to create something more than a disparate catalogue of people's work. At Kyoto, that is exactly what happened. There were several hundred sessions spread over three different venues, so everybody who wanted to organize a session had their 15 minutes of fame. Unfortunately there seemed to be no attempt to put those together into a concise exposition of current major issues, let alone agreed actions. Our Japanese hosts were very hospitable and the organization was formidably efficient, but the coherence and purpose were missing.

What were the big subjects? I had hoped that the role of local government would be prominent, after it had become belatedly recognized at Bonn. But it was hardly in evidence. Sanitation and hygiene were well-covered, but the crucial debate on balancing the water demands of agriculture and the environment was not prominent. Surely globalization and privatization would provide some lively discussion? In the event, however, they only attracted protest. I was pleased to see so many participants from the social justice movement around the world at Kyoto, but saddened that most of them had more interest in shouting slogans than in listening to other people's views, which just alienated the majority of the delegates.

The one headline event at Kyoto was the launch of the report of the World Panel on Financing Water Infrastructure, chaired by Michel Camdessus, former Managing Director of the IMF. I welcomed the idea that financial experts should scrutinize the funding aspects of the water sector, because for a long time we have talked among ourselves. Unfortunately too much of the report was devoted to telling us things we already knew; I would have preferred the panel to concentrate on what they, in their professional work, could do to improve the macroeconomic conditions in which water development can occur (such as correcting global inequities in agricultural subsidies and in trading, which keep poor countries poor).

The World Water Development Report was also launched at Kyoto. This was a massive tome that had all the hallmarks of production by a committee. Its editor had the tough job of breathing life and interest into a compendium of material contributed by no fewer than 24 UN organizations. While some of the material itself was excellent, I personally found it less readable than Peter Gleick's biennial World Water reports, which probably cost a small fraction of the money spent on the WWDR.

The now-obligatory Ministerial Declaration was more feeble than usual, reportedly because the Forum hosts wanted to avoid controversy so the wording was watered down constantly. For example, the delegates did not seem to realize that the UN itself had already confirmed that water is a human right, so the Kyoto Declaration failed ignominiously to acknowledge this. This was hardly an edifying example of political leadership on water.

4.9 Closing Thoughts

Kyoto marked the anticlimactic end of a busy sequence of meetings over 6 years. Alongside those I have mentioned here, the annual Stockholm Water Symposium was expanding during the same period from a purely technical to a policy meeting and becoming a compulsory event for the professional conference attendants. I know that we held more conferences than necessary to agree the policies and targets. They have been validly criticized for their unclear purpose, excessive size, duplication and cost, and yet I feel privileged to have attended them and been part of the process of global policy development.

Now we are trying to concentrate on getting the work done. Meanwhile the World Water Council and Government of Mexico have planned the Fourth World Water Forum, and there is considerable apprehension as to its purpose, cost and usefulness. I hope that in future the size and frequency of big global meetings will decrease and we can make better use of small regional and sectoral meetings to share lessons and monitor our progress.

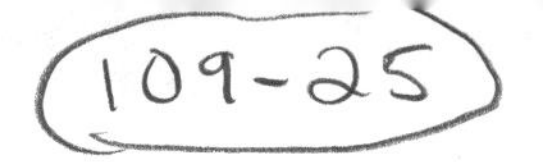

5 International Water Conferences and Water Sector Reform: A Different Approach

Anthony Milburn

5.1 Approach

This chapter is offered as a discussion piece. It is not only concerned with looking back at the successes and failures of the megaconferences. It sees these as part of a process, observable in the latter part of the 20th century, of a huge expansion of information generation and exchange. The analysis notes the other players/actors in the water sector, besides the conferences. It notes the apparent failure of all actors combined, including the conferences, to secure the essential and more timely reform of the institutions for the management of water, and the frustration this has caused within the body of water professionals, particularly those working at international level, and others. The failings of the sector to better articulate and advocate its arguments are examined. An outline of an alternative, comprehensive and tightly focussed process of critical analysis of the sector and its needed institutional reform is suggested, designed to overcome the identified deficiencies.

This process would seek to discern, among other factors:

1. Are the water sector's claims and assertions as central to development and ecological preservation as it claims; or is it some way down the pecking order of global challenges?
2. If the sector's claims and assertions are valid, what is preventing it from securing the resources and commitment to the reform it seeks?

The process would lead on to the production of a do-able plan to reorganize the management of water, to overcome the current institutional deficiencies, to provide the essential protection of natural ecosystems and the interests of future generations and to realize the great potential of effective water resources management.

5.2 Introduction

The Bangkok workshop was tasked with reviewing the effectiveness of the global megaconferences held in the latter part of the 20th century, particularly with regard to their impact on the water sector. One of the motivations for this is a clear and strong sense of frustration among many leaders in the sector that, despite all the talk, the much needed worldwide reforms of water management attitudes, behaviours and practices are still very slow in coming.

This chapter will argue that the megaconferences have been only one part of a huge expansion of developing, articulating and communicating new and more relevant approaches to managing water, which has resulted in an expansion of understanding in the ways experts perceive the water management challenge. However, this has now to be converted into more timely and relevant actions to reform and substantially improve the institutions for water management, and the global megaconferences are not the way to do this. A very different approach is needed, in the form of a more tightly focussed process, which is outlined later in the chapter.

At this point a few statements on what this chapter is, and is not, would be appropriate:

- It is not a blueprint for future action. Rather it is a discussion piece which has been produced for the Bangkok workshop.
- It is not based on specific, widespread and meticulous research. Instead it is based on intuitive deductions from listening to many presentations at many events, including some of the water megaconferences; personal discussions with a large number of people in the water sector from around the world, from the most exalted to the very humble; a wide reading of literature, not just restricted to water, etc.
- It is based on a perceived need to consider whether a different approach to the growing challenge of more efficient water management would help bring about the needed change.
- The views expressed are very much the personal ones of the author, although they will probably find widespread resonance in many quarters.

5.3 The Latter Part of the 20th Century: The Information Explosion

The second half of the last century was characterized by a massive expansion in information generation and exchange. This was driven by a number of factors, among which were:

- New, more sophisticated and more rapid means of electronic communications, surveillance, analysis, etc., of many kinds.
- A major expansion in international travel, driven in part by much more accessible air travel at significantly lower costs.
- A big expansion in tertiary education and research and the growth of Scientific Technical and Medical (STM) publishing (e.g. journals, books, reports, etc.).

There are other factors but the main purpose of these few sentences is to highlight the fact that, as part of this process, new insights and thinking on water management, *inter alia*, were developed, discussed and articulated in a wide variety of media and fora. Many actors played key roles in this, not least the national and international professional associations involved in water. The range and scope of the evolving water management expertise was wide and varied – some was local, the other being of major global strategic importance.

5.4 Reflections on the Megaconferences

A number of general observations can be made:

- The series of water and environmental conferences since the Mar del Plata Water Conference of 1977, together with other important initiatives and activities (of which more later), have contributed towards raising the profile of freshwater and its associated challenges, helping to put water high on the current political agenda. They have helped to draw attention to the complexities inherent in the successful management of water – technical, social, political and economic. They may have contributed to the motivation of a few countries to reform their water management practices, although this is not easy to prove.
- They have provided a platform on which key principles of water management could be articulated and widely disseminated; and the need to approach water management in an integrated, holistic way could be emphasized.
- Those water and other conferences which have been organized with flair and imagination, and for which there has been a significant investment in promotion and in publicity, also carried out with flair, have been the most effective in spreading their message, in making strong points to decision-makers and in educating some of the public at large on the issues, to the extent that some beneficial change has resulted. Such conferences have also put a world spotlight on specific issues and given the advocates/champions/proponents of those issues an enhanced exposure to the media and to some decision-makers, and thus better prospects to influence key political and other decision-making processes. It is debatable, however, just how effective the water sector has been in this, compared with other parts of the environment.
- The big water conferences have attracted a lot of people and official representatives of many countries. However, it is questionable whether the real power brokers in a country, e.g. Ministers of Finance and/or Planning, have ever attended or were particularly influenced by the conference outcomes. It is known from the research how poor is the water sector in many countries in arguing its case for a greater share of the "investment pie", and for greater resource mobilization to tackle its problems. Not least the problem is continued fragmented sectoral approaches and weak advocacy skills. One senses that the water conference attendees largely talk to themselves and, however influential they may be in the water sector, still have problems to influence their national and regional administrations on their return home.

- The ministerial meetings associated with the World Water Forums have been, by and large, disappointing. Their outputs have been mostly generalizations which have not involved any commitment to acceleration of the water management reform process. Despite the excitement of some water professionals at gaining access to these meetings in order to try to influence ministerial thinking, the results have not been particularly productive.
- A lot of the content of the big conferences is not original, nor is it of universally high quality, involving much rehashing of existing knowledge and argument. Where high quality sectoral information has been presented, it has not been synthesized into a holistic, integrated paradigm for the improved management of water. Rather we have witnessed the continued repetition of the IWRM mantra, without benefiting from insights into the improved ways and nuances in which this can be applied in practice; or indeed giving due consideration to the many practical problems of implementing the IWRM approach.

One can summarize the impact of the large conferences in a number of short and pithy statements:

- They have played a part, but only a limited part, in a major expansion of our consciousness worldwide of the freshwater problems and challenges.
- They have contributed to mankind's current position in which we now know most of what we need to know to manage water better but are poor at implementing it and in passing on our know-how.
- They have clearly reached the point of diminishing return in terms of their contribution to water sector institutional reform. Thus, a different approach is needed.

5.5 Other Important Actors

Towards the end of the last century, in the final 10–15 years, a number of new players emerged onto the international water stage and new alliances were forged between pre-existing players. Mention has already been made of the professional water associations, both national and international. A number of these have become more assertive in international fora and there has been some, limited, realignment within them. Among these, the creation of the International Water Association (IWA) from the merger of International Association on Water Quality (IAWQ) and International Water Supply Association (IWSA) is noteworthy. The instigation of the annual Stockholm Water Prize, with the associated Stockholm Water Symposium, was a key innovation at the end of the 1980s. Strongly supported by senior United Nations figures, parts of big industry, the professional water associations and others, it has become the main annual event in the international water calendar for new thinking in scientific/technical, management, policy-making etc., areas of water. The mid-1990s saw the creation of the World Water Council (WWC) and the Global Water Partnership (GWP), the former as a water think tank, the latter dedicated to the improvement of water management skills.

These and many other initiatives, such as PC-CP, World Commission on Dams etc., have all played their part, alongside the existing organizations and institutions, in the development of thinking on improved water management.

The growth in the publication of a wide range of journals, books and reports on many aspects of water has contributed greatly to the expansion of understanding and know-how. The many conferences and other interactions within the professional associations active in the water sector have played a major role. The work of the multi-lateral and bilateral agencies active in the water sector have also helped to spread new awareness and appreciation of the problems, challenges and needed solutions within the sector.

However, it does have to be said, emphatically, that despite the proliferation of organizations and knowledge, mankind is still far from tackling, with sufficient urgency, the vitally needed reform of its water management practices. Just how far this is a failure of vision, advocacy and leadership in the sector is examined in the next section.

5.6 A Failure of Vision, of Advocacy, or of Leadership: Or All Three?

There is clearly a high level of frustration among some prominent international specialists in the water sector about what they perceive as the slow pace of water reform. There are some (Guerquin et al. 2003) who assert that the needed reform is under way but that the reform process involves such fundamental cultural shifts in thinking that it is, of necessity, a slow process. However, it is difficult to prove or disprove this statement. What is plain is that, compared with, say, trade in endangered species or the Montreal protocols for the stratosphere, the pace of the necessary reforms of water is far slower. Thus, it could be argued that the water sector has either failed to make its case to those who matter, or that it does not have a strong enough case compared with other key factors in development and security.

The sector does not seem to produce such effective champions (individuals or organizations) as other environmental sectors, e.g. biodiversity, seas, climate, forests, endangered species, stratosphere, climate, etc. Witness the fact that almost every part of the environment now has some sort of major convention, treaty, protocol or similar, which is well publicized, whereas water has very little comparable legal instruments. Likewise, many members of the public are knowledgeable about other environmental sectors and committed to their protection, whereas this is not so apparent with water. Even in the quality media, which one would expect to be well informed about the issues, misconceptions persist. There is a sense that the water sector does not attract the sort of outstanding communicators, intellects and campaigners, compared to other sectors.

Leading figures in the water industry allege that the world is heading towards a water crisis but the specifics of this, together with the consequences, are not well spelled out. The sector also alleges that provision of safe and secure water supplies is essential for poverty alleviation and socio-economic development, but has not done well to quantify just what this means. The sort of arguments advanced has been on general humanitarian and ecological grounds, and have tended more to emotional appeal rather than hard facts. Further, the arguments are not well enough supported by hard, valid, accurate and detailed data and statistics on the tangible benefits of improved water management – both the positive ones, in the sense of increased prosperity, and the avoidance of the negative impacts of major ecosystem destruction.

Generally speaking, the water sector is not quick to identify and implement potentially relevant new technology. Much of the very large amount of water research which is carried out is repetitive, at times lacking relevance and rarely focussed on the crucial water problems of the lower income countries. Where other sectors have been quick to adopt modern techniques of forecasting and future scenario development, the water sector has lagged behind. Granted that there may be some controversy over some of these techniques, they do nevertheless help to identify significant concerns which require further investigation at least, and sometimes much more. Little of the work done on climate predictions has yet found its way into widespread usage in the water sector, despite the obvious implications for the sector of potential climate change.

There is clear potential within the sector for a significant increase in entrepreneurial activity, particularly in the area of expanding water service provision. Yet the sector is slow to see the benefits and to identify workable ways to create an appropriate enabling environment. Arguments about the pros and cons of private sector involvement become polarized into extreme views, and advocacy of sensible arguments for appropriate commercial activity are drowned out. The enormous potential of a much improved gender balance in water management is very slow to be realized. The vast amount of available, very good and relevant information and expertise on how to solve problems in all parts of the sector, too frequently languishes unused in the literature, if it is even published at all. A major effort to collect, codify and disseminate the needed information and training would bring big improvements to lower income countries. A number of initiatives have been proposed but have not come to fruition.

Water management policies and strategies have for too long embodied crucial exclusions—they have excluded many of the world's poor from water and sanitation service provision, from adequate water to grow the food they need and for livelihood purposes, from protection from water related hazards, from energy provision from hydro-electric sources and many others. These policies and strategies have also excluded adequate consideration of the world's many ecosystems, by diverting and reducing crucial environment flows, by increasing the burden of pollution from communities, industry and agricultural activities, etc. The net effect is a serious and continuing reduction in the biodiversity of freshwater species and a consequent loss of vital resilience of natural ecosystems. Also excluded from consideration have been the interests of future generations.

There is now a slow but growing realization that these policies of exclusion must be reversed and an inclusive, sustainable approach be adopted instead. Only slowly is it being accepted that the poor are not a liability but an asset in a potentially wealthier and more abundant world. Equally slow is the realization of just how vital and irreplaceable is the vast array of essential services which natural ecosystems provide to humanity and thus the crucial need to preserve and protect them for future generations.

Moving to an inclusive and sustainable approach, which would fully embrace the needs of the poor, the environment and future generations, requires a major shift in values within the water sector worldwide. It is essential that this shift takes place. For the culture of the institutions through which water is managed, is underpinned and informed by values. And the needed institutional reforms will be ineffective unless accompanied by a new and more appropriate set of fundamental values, based on an inclusive approach.

It is asserted regularly that reliable supplies of water and access to sanitation are essential for poverty alleviation and socio-economic development, for current and projected populations. Expansion of water resources appropriation to ensure adequate provision of food for a growing world population will be required. Presumably, increased food production will reduce already crucially low environmental flows in critical areas, and increase pollution from a growth in the use of agricultural chemicals. Once development has been kick-started in poor areas, it will most likely set in motion a build-up of small-scale industrial activity, which past and present experience suggests will cause severe water pollution. Despite all of this potential threat to the security of the world's water resources, quantitatively and qualitatively, there is little sign that the likely problems are being properly anticipated and provision being made to alleviate them. Rather the attitude seems to be to expand water use (and likely abuse) and deal with the consequences later. Since there are many who allege that we are closing in fast on the limits of the Earth's carrying capacity, particularly in respect of sinks for pollution, this seems a dangerously blinkered approach.

Environmentalists have done a good job of pointing out the growing problems of water pollution, reduced environmental flows and loss of crucial freshwater biodiversity, with consequent loss of essential ecosystem resilience. Ecologists associated with the water sector have argued for a much stronger emphasis on an ecosystem-based approach to water management, but their success in persuading governments to adopt this have been limited.

Anyone attending the major water conferences, or reading key water literature, would quickly become aware of the seeming complexity of the challenges of water management. As many water experts from advanced countries have noted, applying the recommended tenets of contemporary integrated water resources management is a difficult and challenging task. Countries involved in the application of the EU Water Framework Directive will vouch for this fact. Despite this, ever more complicated additions to the IWRM approach are proposed. Thus it is no great surprise that many lower income countries are struggling to apply IWRM approaches to water management and to establish water management regimes appropriate to their particular stage of development. On the basis of the foregoing, it is of no surprise that water does not figure prominently in most Poverty Reduction Strategy Papers (PRSPs).

Thus a number of crucial questions need to be addressed:

- Is the recommended IWRM approach becoming too complex and theoretical for all but the most advanced countries?
- If so, what would be a more appropriate approach for lower income countries without compromising too much the sustainability interests of present and future generations?
- Is there a possible 'tiered approach' to IWRM which initially could be simpler to apply sustainably, for countries with skill and capacity problems, but which could be expanded and upgraded progressively as skills are acquired, development is accelerated and capacity developed?

Not all of the problems in the sector can be laid at the feet of the people working within the various sub-sectors of water. Clearly there are major cultural and general attitudinal factors and widespread governance problems involved. Yet there is a strong sense of too much conservatism, of complacency, maybe too of frustration and helplessness, and of a failure to look far enough ahead to anticipate major problems which could be avoided, of a failure to adequately make the case for the needed resources to realize the full potential of comprehensive and effective water management.

5.7 Where Might the Leadership Come From?

To what extent is all of the above a failure of leadership? It seems fair to say that the water sector has not, to date, produced outstanding leaders or communicators, compared with other environmental sectors. In the early days of the final quarter of the 20th century, the sector looked to the United Nations for leadership. The United Nations Mar del Plata Water Conference of 1977 suggested that this leadership might be forthcoming, but the momentum was not maintained. A major factor is that freshwater interests are dispersed among a significant number of United Nations agencies with no strong central entity to pull all of these interests together and provide strong leadership to take the sector's interests forward more quickly. Since control and authority over water resources rests with governments, the United Nations' inability to influence more positively the actions of the member countries is a disappointment.

Cynics might argue that since the wealthier nations do not seem to feel that their strategic interests are particularly at risk from the water problems of poorer countries, the needed energy, direction, leadership and resources to accelerate beneficial change are not being provided. Most wealthy countries are relatively complacent about their water problems, figuring that they have the needed know-how and resources to deal with problems. The elites of many lower income countries seem almost as complacent. Perhaps, as Stephan Schmidheiny has observed (Schmidheiny 1992) ".... The painful truth is that the present is a relatively comfortable place for those who have reached positions of mainstream political...leadership".

Statistics on poverty and child morbidity/mortality, so much of which could be prevented by improved water management, seem to have little impact. Even the major world religions, for all their professed concerns for the poor and underprivileged, appear to have had little influence on the conscience of large swathes of significant public opinion in respect of the crucial importance of water.

Expectations of strong leadership and advocacy from the World Water Council (WWC) and Global Water Partnership (GWP) were raised among some sector experts when these were launched. However, such expectations were probably too high given the very limited resources available to these new organizations. The WWC's Vision for the future of the water sector attracted interest at the time but has not, so far, apparently persuaded lower income countries of the need for faster water reform. The GWP has done, and is doing, valuable work, but this tends to be at the level of technical and other aspects of water management expertise. Neither organization has yet succeeded, it would appear, in providing the leadership and motivation to speed up water reform to the required rate.

Perhaps setting up two organizations, rather than one, perpetuated the notion that the water sector is fragmented and added to their difficulty in attracting the resources that are needed to provide the needed visionary leadership for the sector. In addition to the WWC and GWP, many other initiatives, too numerous to list here, have been launched. Most, if not all, are very worthy, as indeed are the people who work in the water sector. All are contributing parts of the jigsaw of how to improve water management. But the truth is that the strong central idea or argument, the outstanding leader(s) and communicator(s) are lacking, to the extent that needed water management reform is lagging way behind the needs of the poor, the environment and the achievement of the Millennium Development Goals (MDGs). That is, of course, unless the water sector's assertions about the problems of the sector are not as urgent as it alleges or certainly not as important as the problems of other sectors, all of which compete for scarce resources.

5.8 The Political Spotlight Is on Water: But Will It Waste the Opportunity?

A classic example of the water sector's problems is provided by the failure to cash in on one of the key issues at the Johannesburg WSSD. United Nations Secretary General, Kofi Annan, identified the five WEHAB (water, energy, health, agriculture, biodiversity) sectors as central to poverty alleviation and socio-economic development. Each of these sectors is heavily dependent, in one way or another, on freshwater. A well-organized and coordinated water sector would have seized on this heaven sent opportunity to strongly assert the case for initiatives and resources to accelerate the water reform process. However, after a brief flurry of interest, the impact of WEHAB seems to be diminishing.

The water sector regularly asserts that safe, reliable water supply is essential for alleviating poverty and furthering socio-economic development. Belatedly, some economic data on this is now appearing. But it is limited in scope and does not give an impressive global picture of the benefits to all, rich and poor, of the advantages of access to water for all, including the natural environment. A more convincing set of arguments would give powerfully convincing data on the beneficial impacts on increased global GNP, the benefits of enhanced trade for richer and poorer nations, the big health improvements, the alleviation of the threats of major ecosystem collapse and the greater security which would come from removing poverty. The richer countries might not be swayed by images and data on sick and dying children, but make a convincing case about enhanced prosperity and less disaffected minorities/greater first world security and attitudes might begin to change.

To show that some of these issues are not far at present from some Western leaders' minds, consider the following words of the former UK Prime Minister Tony Blair (Blair 2005) in an issue of *The Economist*. Speaking about the UK's forthcoming presidency of the G8 countries, he asserted that the problems of Africa and climate change will be major agenda items. But among the points he makes are some which are germane to the water sector's interests. The following quotations seem relevant:

> Should this [the African problem] matter to the rest of the world? For democratic governments, it should, because it matters to our citizens. They give millions of dollars in aid to the countries involved and their people. They campaign for their governments to do more. They passionately believe, as I do, that it cannot be morally right, in a world growing more prosperous and healthier by the year, that one in six African children die before their 5th birthday. The worldwide campaign to make poverty history rightly challenges us to act.
>
> .... We must now all accept the utter futility of trying to shut our borders to problems abroad. Famine [in Africa] will affect our countries because it will be a trigger for mass migration. Conflict, too, drives millions to flee their homes. Both create the conditions for terrorism and fanaticism to take root and spread directly to Europe, to North Africa and to Asia.

On climate change, Blair states "Finally, some argue that there are more immediate problems. In some senses they are right. Over the next 5 years, e. g., water pollution will cause more harm worldwide. It is wrong, however, to see these problems as mutually exclusive. Without a stable climate, addressing other environmental threats will be impossible, ensuring a future of more degraded water and land."

In the latest 'State of the World' report (World Watch Institute 2005), looking at issues of population and security, the authors note that four demographic risk factors – growing proportions of young people, the HIV/AIDS crisis (which is accompanied by increased susceptibility to water-related disease), rapid urbanization (which has big water implications) and reduced availability of cropland or water – can, singly or more often in combination, lead to domestic instability and even international insecurity. Clearly from this, water is central to concerns about security.

These statements are included in this chapter because they point to both the opportunities and the threats facing the interests of the water sector. Assuming that Mr Blair will carry his assertions to the G8 meetings, and making due allowance for political rhetoric, opportunities reside in his recognition of the scandal of poor countries – child mortality (much of which is caused by water-related diseases), famine and the threat to western security and interests of mass migration, terrorism and fanaticism. His precise views on the relative threats of climate instability and water pollution are not so straightforward to discern but he seems to opt for climate stabilization as his principal concern. Given that it would take decades of concerted effort to achieve such stability, even assuming mankind possesses the needed ability to do this, set against the mounting threat from water pollution over the next two decades, ought to be cause for concern for the water sector. The 'State of the World' report points directly to the need to ensure adequate and reliable supplies of water as a major factor in ensuring security.

Given the influence exerted and wielded by the G8 countries and the State of the World reports, one can ponder how the water sector has responded to the statements made. Probably it has not responded specifically at all, which emphasizes the point made earlier that the sector is weak in asserting its case for greater attention and resources.

5.9 Moving from "What Could" and "What Should Be" to "What Is Going to Be"

The megaconferences have been characterized by a multitude of statements on "what could" and "what should" be done to improve water management. Frankly, the world has had enough of these. The water sector has, by and large, the knowledge it needs to radically improve the way water is managed. Certainly some of this know-how will need refining in the light of widespread applications in practical situations. But enough is known about the essentials to make a much better fist of the management process than at present. Thus, the challenge is clearly about how to move to the "what is to be done" phase, via a set of feasible plans for the accelerated reform of how water is managed.

Water professionals' assertions on the centrality of water in poverty alleviation and socio-economic development and the pivotal role of water in the five WEHAB sectors seem not to have convinced a significant number of governments. There is no firm evidence that the UK's Mr Blair is convinced of the centrality of the water argument. Nor, from the evidence of a significant number of lower income country PRSPs, are many governments in countries which have prepared such papers. This situation gives rise to a number of crucial questions:

- Does the water sector have a strong case which it has failed to argue convincingly enough?

- Are its claims not strong enough when set against the demands of other key factors in the poverty alleviation/socio-economic development process?
- Is its case not couched in the best terms or not using the best and most persuasive arguments to convince decision-makers?
- Is the water management reform process proceeding at the most feasible pace, commensurate with the resources, capacities and perceived priorities of the lower income countries?

Probably, the truth lies somewhere between these different scenarios. However, for its own peace of mind, the sector needs to examine these and other related issues most carefully to help it determine the best possible future strategy.

To date, the group of prominent water specialists and their supporters who set up the WWC, GWP, Stockholm Water Symposia/Prize and the many other initiatives in the sector have done a superb job of raising the political profile of water issues to the highest international levels. They have done this with very little resources relative to the successes they have achieved.

They have also raised understanding, over a much wider scale, of the challenges of managing water holistically, the techniques which can be used to do this, and many of the issues which have to be addressed. They have worked to condense out many of the actions which need to be taken and, in general, how to do these more effectively than before.

The challenge is how to take the process on from here. The political spotlight is currently on water. But a key question is whether it can be kept there long enough to secure widespread long-term commitment and resources to speed up the rate of beneficial change and reform. Failing this, there is the risk that some other global challenge will come along which pushes water out of the spotlight and back into the shadows, until major disasters force the needed changes.

A key task involves mobilizing very large resources and managing these effectively and efficiently to bring about the needed faster rate of change. It is also about changing attitudes, perceptions and behaviour in a wide range of cultural settings, to persuade diverse governments to engage themselves and their countries in accelerated water reforms. Given the size of the challenge and the resources required, the stakes are high.

5.10 It Is a Big Job: But Others Have Done It

All of this tends to argue for a different approach altogether from the one followed so far, especially that of the megaconferences. The challenge is not unlike that faced by a global corporation confronted with the need for major strategic change or about to embark on a significant new enterprise. The thoroughness of their approach is instructive – it has to be since very large sums of money are at stake, with critical shareholders looking over the company directors' shoulders. Large egos are at stake too, who hate to get things wrong. Such companies adopt a process which will test very thoroughly any assumptions they may be tempted to make; they research the background of all aspects of the initiative they are considering and apply the highest possible standards of critical analysis and interpretation to data collected and assumptions made. They will do this via a combination of experts and generalists within the company, augmented by high quality external expertise, organized into task forces focussed on the critical issues. The results of all of this come back to the Board of Directors who have to make a collective judgement about whether, and if so how, to proceed.

There are strong arguments, to the extent that this is feasible, for the water sector to embark on a comparable process. An outline of what this could be is set out below. It is not intended that what follows is definitive. Rather, it is illustrative of a process which could be followed.

In essence this process entails critical examination and analysis of crucial cross sectoral issues, cross-sectoral processes and sub-sectoral issues and processes, followed by the synthesis of the conclusions into a plan for action. Some examples are given below to illustrate the thinking.

1. Cross-Sectoral Issues

- What are the central assumptions in the sector and how correct are these?
- Has the sector, collectively, identified the central and critical issues of water institutional reform and, if not, what are they?
- What are the detailed quantified true and full benefits which would accrue from a much improved management of water, e.g. specific impact on global GNP, health, prosperity and abundance, prevention of the poverty driven problems of extremism, terrorism etc.; prevention of ecosystem collapse, preservation of essential biodiversity etc.; and, crucially, other key factors which the sector has yet to identify?
- What factors and/or arguments in the water sector, or legitimately connected with it, would produce sufficient weight of public opinion to put the required pressures on the political process to accelerate reform?
- Just how critical is the world water problem compared with other major environmental and socio-economic and development challenges, taking proper account of the pervasive use of water in so many aspects of life and the increasing interconnectedness of modern living?
- In much more specific terms than hitherto, just what are the overall socio-economic and environmental consequences of maintaining the present slow pace of water reform?
- And many others.

2. Cross-Sectoral Processes

- What are the principal factors in the political economy of reform in lower income countries which are most amenable to influence and manipulation, to secure more rapid water reform?

- What are the true and essential investment needs of the sector which are pivotal to setting in motion and accelerating the process of poverty alleviation/socio-economic development to assist the world's poor; and how can the needed funds feasibly be raised, disbursed and managed effectively at the different levels needed?
- What measures are needed to ensure that – with a major expansion of world population, accompanied by increased urbanization and the provision of improved water services to most if not all – the natural environment will not be overwhelmed in critical areas by the burden of pollution?
- What are the principal institutional functions and services essential to a moderately well-functioning water sector (including sub-sectoral needs). How can institutional deficiencies and capacity constraints most realistically be overcome?
- What are the crucial values which must underpin the institutions active in water management at international, regional and local level?
- What are the most effective models for integrating the many challenges of water's sub-sectors into the national planning and budget process of countries? How should this be developed in terms of policy and plans?
- Given that many countries are experiencing problems in implementing integrated water resources management (IWRM) now, and experts from more advanced countries continue to add layers of complexity, what realistically applicable forms of IWRM can be devised which will encourage faster adoption and application of the key principles?
- Given that it seems that many if not most of the world's water problems have been tackled successfully somewhere, by someone at some time, but that information on this is very badly disseminated, how can the body of available expertise particularly applicable to lower income countries be collected, codified and better disseminated?
- Is the training of water personnel at all levels as relevant to the needs of water management in the 21st century as it should be, or is there the need for a radical overhaul of this?
- Are there new technologies either on the fringes of the water sector or available, tried and tested in other sectors, which could bring very big improvements to the sector, especially in lower income countries?
- Is the vast amount of research currently carried out on the many aspects of water as focussed and effective as it should be? Is it really addressing the central and crucial needs of providing safe and reliable water services to the populations of lower income countries? What outstanding research needs to be carried out as a matter of priority?
- And many others.

3. Sub-sectoral Issues and Processes

- What are the relevant issues to consider, drawn from recent experiences worldwide, to enable successful reform and management of irrigation and water services?

- What are the guiding principles and essential requirements for reform and expansion of water supply and sanitation provision, taking account of managerial, technical, social, macro and micro financial issues?
- And many others.

The aim of the above is to give a flavour of what would be an extremely thorough, in depth, critical analysis of the accurate case for water reform and how it can be done; not as at present, what could or should be done. In the event that the water sector, as a result of this detailed analysis, cannot sustain its current claims and assertions, it will at least have a clear and objective view of where it stands in the wide range of global challenges.

Part of the process must include advice and expertise from those other environmental sectors (e.g. trade in endangered species, stratosphere, climate change, etc.) which have been successful in drawing attention to their problem and securing commitment and resources for beneficial change.

The process would draw on expertise from within the sector and from outside, using the highest levels of intellectual ability, analytical skills and practical experience available. To the extent it is feasible, experts should be drawn from lower income countries, making maximum use of visionary minds and can-do personalities. Expertise is needed in the critical analysis of the issues identified above; in the addressing of major strategic challenge and change; in mobilizing, committing and managing large-scale resources; in pinpointing the supporting research and data needed, together with the analytical and critical thinking skills to make the correct deductions; and in pinpointing and articulating powerful and persuasive arguments to make an unchallengeable case for accelerated water reform.

The intention would be to organize the available expertise into focus groups, orientated towards the issues identified above and others to be added. These groups would report back to a synthesizing body which, in cooperation with the focus groups, would produce either a plan or a campaign for accelerated water sector reform in the form of a "do-able proposition". This would then be tested by submitting it to respected government ministers and development specialists, who are progressive and fair minded, yet critical, to ascertain any crucial flaws. Once the review was complete, the conclusions should be presented to an international conference of known leaders from the water sector, particularly in lower income countries, and others in key positions associated with the principal decision-making centres, relevant to water, worldwide. The intention of this would be to convince these leaders, so that they in turn can carry the message to others, in their countries, regions, river basins, companies, agencies, organizations etc., and convince their followers, so mobilizing an accelerated reform process.

A Steering Group would oversee this process, chaired by a prominent international figure possessed of significant political capital. Supporting him/her should be up to 11 additional members – people of broad international vision, with talents ranging from understanding of the political process in lower income countries, mobilization and management of large natural resource projects, experience of development banking, experience of policy-making, planning, budgeting and management in lower income countries, high levels of communication and leadership skills, etc. The Steering Group would endeavour to represent the key constituencies in the various organizations and institutions in the water sector. It would also usefully include someone with top-level experience of successful attitude and behaviour change in another environmental sector, e.g. trade in endangered species or the Montreal protocol.

A process such as this would not be cheap. To do it thoroughly would involve substantial expenditure, of the order of tens of millions of dollars or more. As an indication of the sort of funding, which such projects can absorb, consider the 1995 allocation of 100 million dollars for investigations for the Nile River Basin Action Plan alone (Nicol 2002).

To set up such a process and manage it through to an acceptable conclusion will require initiative and leadership of a high order. This could come from an agency or group of agencies within the United Nations System with a high profile, substantial resources and experience of this kind of sophisticated exercise. Alternatively, it might be motivated and resourced by the G8 countries and led by an entity within them or established by them. In terms of annual global GNP and overall activity worldwide of all kinds, it would be a modest enough exercise in itself, requiring readily provideable resources, if the political will to carry it out is there. Yet the potential returns from a successful outcome would be enormous.

5.11 The Future of the Water Megaconferences

Theoretically, the concepts outlined above could be adapted to a conference format and could even be incorporated into or substitute for the current trend in global mega water conferences. However it would be nowhere near as effective as a well managed, tightly focussed, stand alone project and there is the real risk that a conference format would so dilute the process that it would be ineffective.

As indicated earlier, the established pattern of the very large water and water-related conferences are no longer useful for moving forward the accelerated reform of water management. Thus these conferences should change radically their format or even cease altogether. However there are very many valuable water conferences, e.g. Stockholm Water Symposium, the regular large and the specialized smaller conferences of the many professional associations and the myriad other topic conferences organized on and around water. All of these make a valuable contribution to our total stock of knowledge on water and water-related issues and should continue.

5.12 Conclusion

The megaconferences have to be seen as part of a broad series of initiatives which have investigated, clarified and articulated the principal challenges in the water sector, foremost among which is poor water management. Yet the timely reforms of water management, to a more efficient and sustainable approach, are slow to materialize. Arguments have been advanced about the central importance of water in poverty alleviation/socio-economic development, security, health, ecosystem maintenance and others. The profile of water has been raised to a high political level, yet the initiatives and investments needed to secure essential and timely water management reform have not yet been secured. The chapter has suggested an alternative process – to clarify the issues, articulate powerful and convincing arguments, produce a do-able plan for improving the management of water resources worldwide, and thus establish clearly the priority claim of water on the world's attention and resources (or otherwise), thereafter to speed up the rate of management reform.

References

Blair T (2005) A Year of Huge Challenges. An Invited Feature Article in The Economist (1 January)

Guerquin F, Ahmed T, Hua M, Ikeda T, Ozbilen V, Schuttelaar M (2003) World Water Actions: Making Water Flow for All. World Water Council, Water action Unit, Earthscan, London, UK

Nicol A (2002) The Dynamics of River Basin Cooperation: The Nile and Okewango Basins. In: Turton A, Ashton P, Cloete, E (eds) Transboundary Rivers, Sovereignty and Development. Green Cross International and University of Pretoria, pp. 167–186

Schmidheiny S (1992) Changing Course: A Global Business Perspective on Development and the Environment. MIT Press, Cambridge, MA

World Watch Institute (2005) The State of the World 2005: Global Security (22nd edition). Worldwatch Institute, Earthscan, London, UK

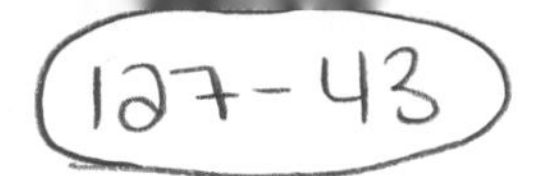

6 Megaconferences: Serious or Circus? An Unscientific Personal View

Gourisankar Ghosh

6.1 Introduction

This analysis is different from the rest as presented in this important book on megaconferences. It is based on a personal impression, analysis and assessment and not based on any statistical analysis or any proper research methodology. It is a personal view of the industry called megaconference. However, having attended several such conferences as a member of government delegation, United Nations agency and also as a representative of partnership organization, I have possibly developed a rounded but also slightly cynical view of the whole process. However, I have attempted to present my views without any bias and prejudice. It captures the experience of a water professional who has attended a number of mega, macro and micro conferences in water and also non-water areas. It is intended to be of some help to researchers or policy-makers to understand some points that they may consider further. But for the continuous encouragement from Asit Biswas this chapter would not have been written. He has been the main mover behind this whole exercise which is very topical and timely, especially before the World Water Forum in March 2006.

I have used some of the findings of recent independent evaluation of the Dakar WASH Forum as organized by the Water Supply and Sanitation Collaborative Council in November 2004. Though not a megaconference, in terms of either its size, attendance or cost, it may still work as a model, results of which can be used in an even larger scale! Before organizing any mega or even macro conference, one should ask a fundamental question as to how much that event will change the lives of millions of people who are waiting for 'us' to create a conducive environment for their access to basic needs such as water, health, sanitation, education and energy.

6.2 Who Organizes such Megaconferences and Why?

For the purpose of this analysis, a megaconference is defined as an event having an attendance of more than 2000 people or a cost of more than $10 million. The definition is arbitrary, but we have to draw a line somewhere. In order for an event to be considered a megaconference, it need not even be a global conference: it can even be regional or local.

Usually, for a megaconference to be a success, it should be organized by a known organization or institution which has a wide network and outreach, or by the United Nations or its agencies or related networks, or it evolves around a topic which has great appeal and can encourage people to participate through its messages, content and possibilities. Besides the main factor for attendance of a megaconference is the presence of a donor or agency funding a number of participants to attend. Almost all the megaconferences follow this principle. Let us look at some of the reasons as to why they do so.

- To respond or to lead to a United Nations resolution.

  Most often a conference is a product of another conference. It has almost a chain effect. The same people behind the initial conference plan and plant the seed for another, either for the purpose of genuine review or just to get the momentum going in terms of consultation. Unfortunately, in most cases, it becomes a routine chain on conferences without any clear goal, objective or direction. The annual Commission of Sustainable Development, as initiated within the United Nations System for the follow up of Rio process and thereafter, now assumes a significant shift from a mere diplomatic talk shop to a forum in two parts, where the first week feeds into the second week of a high-level segment with experiences and knowledge from the practitioners and sharing their experiences with the delegates from the member states. It becomes a mandatory routine for the United Nations agencies to be 'seen' and 'heard' in those conferences. In fact, eyebrows are raised if someone is not present for a genuine and valid reason.

- Interest of a single individual or a real mover and shaker for an organizational interest.

  Sometimes an organization has nothing much to show in its activities, except for organizing one forum to the next. That covers the existence and activities of that organization for the interim period. Such an organization survives and gains or maintaining importance and relevance through organizing the megaconferences. One such example is the World Water Council. Another type is an agency or NGO/IGO such as the International Aids Society created for this purpose only implementing megaconferences on behalf of an international organization like UNAIDS.

  Unfortunately such organizations also develop vested interest groups that are normally well positioned, well informed and influential, and they normally control the planning and designing of the activities and process towards a megaconference which more often than not confuses the real issue. The design and process is mostly not transparent, non-democratic, arbitrary and does not respond to the demand and current need. There is a great deal of discussion during the event on the participation of the stakeholders, or bottom-up approach. However, during the planning of the megaconference, the process is very much controlled by institutions and individuals having specific vested interests.

In contrast, there are cases where one dedicated person can really make all the difference. One of the best examples of a genuine cause-led conference was where one United Nations agency or one person led the crusade for the cause, resulting in a spectacular global head of the state conference which was the United Nations Children Summit in 1990, resulting in the recognition of the rights of children. The mover and shaker was late Jim Grant, the then Executive Director of UNICEF. He travelled to more than 100 countries to meet with the leaders to convince them about the children's need at that time. In contrast, the World Water Forum in the Hague resulted in a divided forum due to a leader who wanted to follow his vested interest and not allow development of a common platform, and rejected a global vision developed by the people.

- Genuine response to develop a common position to a serious crisis or issue.

  Some global conferences are designed especially to highlight a global issue and develop a plan of action. It took years before the HIV/AIDS issues were brought to such a stage. It also took years and a series of conferences around the world to reach Beijing on women's issues. The whole movement for the Health for All was possible owing to the Alma Ata Declaration which is still the bedrock for public health policies.

- Advocacy (both internal and external).

  All the conferences are supposed to be for advocacy, both internal and external. The organizers may use the event to further their own agenda to highlight their current and future activities. The sector specialists are keen to utilize the occasion to advance their own interests before their own leaders. This is often the case for the United Nations agencies. Mostly, the water sector does not receive high priority within their organizations. Accordingly, they support and use the megaconference for the purpose of their internal as well as external advocacy with their own senior management, donors and other partners. The NGO groups also use the event for pushing their agenda and the same is true for the private sector, other institutions and research organizations. Basically the key objectives of such jumbo conferences are networking and advocacy. These do not make a good platform for learning or knowledge dissemination. They are a platform for marketing!

6.3 Who Attends?

Besides the representatives of the host nation and agency, nearly 50% are the participants and the main speakers in most of the conferences. Within that percentage a group of conference hoppers are also a permanent feature in water conferences. They even proudly display their credentials in their CV as a qualification of attending so many megaconferences. Unfortunately, some of them are supported by international agencies on a regular basis. It is difficult to judge their actual contribution to those international events. Interestingly none of them can really claim to have contributed to any substantial change either in policy or in actual delivery in their countries. Most of the water conferences have also become very much a closed family affair. Hardly any other professional group other than the water crowd is seen in such gatherings. It really makes the input very narrow and broad development discussions are avoided in most cases. It is really very amusing for that reason to listen to some presentations, which basically consider research findings of the 1950s and 1960s, but being presented as the new paradigms and obviously asking for funds for substantial additional continuation of such work.

Besides the professional conference hoppers, the other groups of regulars are the professional conference managers, facilitators, chairpersons and different constituency representatives and government representatives mostly from water subsectors. Sometimes they call themselves the Knowledge Managers. The fact is that in the name of knowledge management, they are, basically, knowledge controllers. Their main role is to package knowledge from the South and then repackage it again to sell back to South. They call it knowledge networking where they develop professional matrices for the poor South to study, or to develop a policy framework. Most of these self-styled knowledge dealers do not have any touch with the ground realities, but they use the system and process for their own survival.

Besides, they also organize tokenistic representations of a few grass-roots people who are seemingly lost in the crowd and are paraded in the glory of the success of some donor-funded projects or by some NGOs to demonstrate their issue. The last but not the least important group is the one whose members are not interested in the proceedings but are there only to enhance their own agenda and narrow objectives. The megaconference is really a circus. Instead of being a three-ring circus, it often has multiple rings and there are shows all around. In the middle is the confused genuine audience. It takes the first two days to settle down, followed by a day's good work and then the process of winding up. At the end only the trapeze artists are remembered flying so high in the air!

Of course we should not forget the tourists who use the megaconferences as a good excuse to visit the country.

6.4 Politics of Megaconferences

No megaconference has ever been organized without a hidden objective or a political agenda. Both the sponsors and the host organization have an interest in internal as well as external advocacy. Sometimes these plans even go beyond the conference in terms of space and time. The agenda is also decided on by the interests of the groups within the country or by any agency or United Nations organization. These interests can vary from purely personal agendas to institutional goals.

The politics is guided primarily by the topical situation, demand and any plan to create a new agency or to promote some individual or a group of people. In contrast to a social mobilization, these people are neither leaders nor have they gained their positions through their technical brilliance. Most often, the good water professionals are sidelined, unless they are specifically backed by some organization or United Nations agency. In recent years, we have also seen the emergence of water leaders who have never even seen a water project in their entire life. This is the beauty of megaconferences: they can create giants out of nothing!

6.5 Difference of Opinions Between North and South

It is interesting to see the divide between the expectations of the participants from the Global North and South. The Global North is normally represented by the experts, researchers, developmental practitioners, United Nations agencies and related organizations. Their expectations are always different from that of South, though both their expectations are supposed to converge for a common cause. The Global North uses the forum as their selling platform, to get endorsement for more support for their programmes and projects in the South, an endorsement of their own activities and overall selling of themselves, besides networking. They always prefer to have the conference structured around technical issues, on themes which may even be obsolete in the current situation but which suits them well to justify their existence. It is in direct contrast with most of the southern representatives.

Most of the southerners are supported to participate in the conference through the projects supported by the Global North. But they want their voices to be heard in the global platform. For them, the conference is a political-socio-techno-economic platform for raising their unheard voice in the international arena and to share their real experience with the rest. They speak not out of the desktop research matrices but from their knowledge, taking into consideration the real variables in life and politics. These contrasting approaches make the conference lively and sometimes conflicting. The demonstration at the Kyoto World Water Forum against the Camdessus report and the discussions on finance is an example where both the presenters, organizers and demonstrators made no attempt for either converging their views or listening to each others' points. Both find such a megaconference platform useful to vent their own views in parallel.

In an independent evaluation of a smaller conference of the Global WASH Forum, nearly all the Global South appreciated the semi-political format of the conference as they found their voices to be heard by the ministers and other leaders, and they could interact with them as equals and also as partners. In sharp contrast the Global North, consisting of mostly researchers and experts, could not mix with the political crowd so easily. They were out of their depth in their appreciation of the policy and institutional issues. One of the most important presentations on sanitation in the plenary was made by a minister who actually started the project, implemented it himself and scaled it up to cover nearly 20 million people. As the effort was without any coalition with an international agency or programme, it did not get much recognition from them. This clearly brings out the problem we face today, where the global aid is mainly filtered through these institutions and agencies and does not reach the poor directly, promptly and easily. The Global North industry for the development is actually being a hindrance to capacity building and empowerment. It is also clear that megaconferences somehow bring both North and South together, and yet so far has failed to develop a joint force for change. It is not merely a suspicion but a divergence of the approach to the ultimate goals which makes the difference. This is the biggest failure of the megaconferences. The reason is primarily because those conferences are often planned by the Global North only. One of the visible exceptions was the Bonn Ministerial Water Conference, organized by the German government, which broke through the barrier and was successful in establishing a good process, primarily because the organizers actually listened to the global advisory group for designing this conference and were serious to the cause.

6.6 Water Conferences: The Usual Outcome

The Global Water Conference's outcome is very confusing. It is like a bubble with a rainbow in it. It looks nice over a short and transient period but does not leave a lasting impression at all. The outputs are too many, sometimes elementary and other times conflicting. It does not resolve conflicts such as domestic versus agriculture water uses and never tried to resolve such issues. Normally it gives mixed and often conflicting messages in abundance, without any consensus on funds, actions and strategies. Most of the presentations are not intellectually stimulating, because of their poor qualities, paucity of time or chaining of too many papers within a short period. There is an attempt sometimes to create a scare like the global water crisis and it backfires. In fact, the water people are very poor communicators and cannot give simple messages to the larger world, and so they lose in selling the sectoral issues. Hardly any new idea or issue emerges in these fora. Often there is an attempt to package old recipes that have not worked in the past, but never enough review of the causes of the failures and the constraints. One learns much more during a smaller conference.

6.7 What Do Others Do?

If we look at the other sectors, except for the HIV/AIDS Conference, there are no regular megaconferences organized as in the water sector. The Health for All: Alma-Ata Declaration (1978) is regularly followed during the WHO Health Assembly. There is no attempt to repeat the global conference. In the case of population, there have been only two major events: Bucharest (1974) and Cairo (1994). The women's issues were developed through a series in Mexico City (1975), Copenhagen (1980), Nairobi (1985), Beijing (1995) and now after 10 years in

New York as Beijing+10: significantly reduced in size and in a review format. The famous and successful Children Summit in New York in 1990 was followed after 10 years in a special United Nations General Assembly in 2002. The Education for All was known still as the Jomtien (1990) and Dakar (2000). The annual Davos World Economic Forum is an upmarket forum, but not a megaconference. The Rio Conference was followed by Johannesburg but not exactly as a Rio+10, but as a stand-alone megaconference based on Agenda 21. It was prepared with great planning for the WEHAB agenda and was fairly successful in drawing attention but failed to get a common agenda and consensus. The follow-up of the WSSD in Johannesburg is still based on the Millennium Goals as decided in the General Assembly in 2000. Not surprisingly, the people remember the MDGs but not the Johannesburg Programme of Implementation (JPOI). A fundamental question that needs to be asked is that besides the addition of the sanitation goal, was there any need for the megaconference?

However, the lesson learnt from this list is that there is *no* need to have conferences on one issue at regular intervals unless there is a clear need for follow-up and a mechanism to judge the impacts and the changes. Is it good or desirable to have a world gathering every 3 or 4 years, only to slap each others' back on the success of the events, irrespective of their outputs and impacts? Is it not necessary to objectively assess if a megaconference has produced any real changes and, if there are some, how they have been achieved, who has benefited and at what cost? Too many events at regular intervals contribute to conference fatigue and lose their advocacy power. It becomes a roaming circus which comes back in the circus season and at a regular interval, with almost the same artists performing the same acts. Only the venue changes!

6.8 Local Challenges with Global Answers?

Despite the attempt to globalize the water and sanitation issues, it is a very local issue and even sometimes localized issue. There can be some global experiences and solutions but the solution must be appropriate to the local situation, local problem and the local institutions. The global platform does not provide the answers to the local problems and it is a paradox that a global solution is often sought for a problem whose solution lies in local knowledge. This is the biggest failure of a global megaconference which is even visible in much smaller ones such as the Global WASH Forum. It is very difficult for the organizers to sit on judgement and eliminate the different themes, solutions and opportunities which are raised in such a conference and to make it satisfying to all.

It is better to label these fora as marketplaces where the best-seller gets attention and sells their commodity at a price. When there is a price, there will be a genuine demand.

6.9 Dakar Global WASH Forum: A Case Study

Dakar Global WASH Forum, as organized by WSSCC, is a macro forum with less than 1000 participants and a total budget of around $1.2 million. Yet, we as the organizers were concerned about its cost in relation to the impacts and asked an independent organization (WELL) to conduct an objective evaluation. The evaluation is very relevant to a megaconference, the cost of which may be almost 30 times this budget and the attendance is much higher and yet the impacts appear to be much less than a macro forum.

The methodology and the results are also very relevant to the concept of the megaconference and possibly strengthen the need for more smaller and regional conferences to raise the issues close to the locations in which they arise, and not take a global overview. The extract from this evaluation is considered next.

6.9.1 Conceptual Framework

The initial step taken in the evaluation was to develop a conceptual framework. The reason for developing this was to:

- summarize and organize the essential elements of the evaluation; and
- clarify the questions that need to be asked.

The OECD Development Assistance Committee (DAC) evaluation template was used for this exercise. The DAC framework outlines the basic structure of a development-related evaluation. Such a framework will normally have sections on effectiveness, efficiency, impact, relevance and sustainability. We adopted this structure, and for each of these elements (apart from efficiency), we identified the type of information that we would need to collect in order to make a judgement about each of the aspects outlined in the scope of work.

Efficiency measures the outputs, qualitative and quantitative, in relation to the inputs. This evaluation has not considered the efficiency of the Forum in relation to financial and administrative arrangements; the necessary data are being collected separately through an internal audit by WHO and will be the subject of a separate report.

6.9.2 Data Collection

We constructed a single evaluation instrument for data collection with participants. This was a questionnaire based on the structure of our evaluation framework. The questionnaire consisted of a number of open-ended questions and several attitudinal questions constructed using Likert scales. During the evaluation, the instrument was administered as a questionnaire in the survey, available in French for those who requested it, and as an interview protocol in participant interviews. A parallel interview protocol was constructed to use with WSSCC personnel.

6.9.3 Questionnaire Survey

The aim of the questionnaire survey was to provide information for a general assessment of the issues raised in our Terms of Reference. Our initial agreed sample size for the survey in the Terms of Reference was 100. However, we believed that this number should be increased to maximize the potential for relevant feedback. As such, we distributed the questionnaire electronically to all participants who attended the Global WASH Forum and for whom we had a valid e-mail address (approximately 478 in total). There were no email address details for 37 participants. Further, we conducted a postal survey of all the participants without e-mail addresses and those whose e-mail addresses failed. In all, 68 respondents (13.2%) responded. This was achieved by sending out two reminders to all participants. Table 6.1 presents the distribution of respondents by their functions.

Table 6.1 Distribution of respondents by function

Employer	Number of participants
Policy-Maker	11
Bilateral	3
Multilateral	2
Research	10
Practitioner	2
NGO	33
Consultant	2
Media professional	1
United Nations staff	4
Total	68

A good proportion of respondents (43.5%) represented NGOs. Representatives of the donor community formed 8% of the respondents. Overall, a good spread of respondents across all the stakeholder groups was achieved.

Table 6.2 shows the distribution of respondents in terms of their country of residence. There was similarly a good geographic spread of respondents with 67.6% of the respondents surveyed coming from the Global South.

Table 6.2 Distribution of respondents by country

Country	Frequency	Country	Frequency
Bangladesh	1	Norway	2
Bulgaria	1	Pakistan	1
Burkina Faso	1	Peru	1
Canada	1	Philippines	3
Central African Republic	1	Senegal	10
Colombia	1	Sierra Leone	1
Cote D'Ivoire	2	South Africa	7
Egypt	1	South Asia	1
Ethiopia	1	Sweden	1
Finland	1	Switzerland	5
India	6	Tanzania	3
Israel	1	Uganda	2
Kenya	2	UK	3
Malawi	1	USA	1
Mozambique	2	Worldwide	1
Netherlands	3	Total	68

6.9.4 Participant Interviews

First, participant location was classified according to 'country groups' data and statistics, categorizing them according to 'low-income economies', 'lower- and upper-middle-income economies' and 'high-income economies' (Table 6.3).

Table 6.3 Distribution of respondents by economic country groups

Economic country group	Percentage at forum	Percentage interviewed
High-income	28	38.8
Upper- and middle-income	12	5.6
Low-income	60	55.6

A random sample of interviewees was then selected, which aimed to match the proportions of Forum participants given above. The initial intention was to interview 50 participants. However, a decision was taken after 15 interviews had been conducted that this target should be reassessed, due to the difficulties incurred in arranging interviews at convenient times within the time constraints imposed by the duration of the evaluation. Every effort was made to interview more participants. Two requests were sent to all the participants. In addition, telephone calls were made at random to solicit interviews but the efforts, to a large extent, were futile. There are two main reasons for this. First, many of the participants were away when reminders were sent. We know this by the 'out of office' replies received and through information obtained over the telephone. Second, there was quite simply, interview fatigue.

We obtained an additional three interviews bringing the total to 18. Notwithstanding the difficulties that were encountered in arranging the interviews, our assessment is that the lower than expected number of interviews did not influence quality of findings. This is because there was no advantage to be gained in continuing the interview process beyond the 18 interviews already obtained, as theoretical saturation had been reached. That is, the point had been reached when there was no new substantive data appearing, compared with existing interviews and completed questionnaires. Therefore, the 18 interviews, together with the large amount of data gleaned from the open-ended questions in the questionnaire survey, adequately represented the balance of opinions among the participants.

In addition to the 18 participants interviewed, two members of the WSSCC secretariat were interviewed in depth about their knowledge and experience of the relevance and impacts of the Global WASH Forum, along with their views about the structure, content and organization of the Forum.

6.9.5 Analysis

Separate analyses of the survey and interview data were carried out by using the Statistical Package for Social Sciences (SPSS™), a software package used to codify and analyse the survey data. For the attitudinal questions in the survey, we tabulated frequencies and relied upon the modal value for reporting, after triangulating its validity by calculating a weighted average for each scale. The interview data were analysed using the techniques of coding, patterning and counting.

6.9.6 Key Findings

A summary of the key factual findings as reported by our respondents is given next.

Sustainability of the Global WASH Forum

Should it carry on?

The consensus view is that the Global WASH Forum should continue to be held. However, it should take the following into consideration.

- Occur less frequently than the previous 2/3 year cycle.
- Emphasize policy dialogue.
- Have greater representation from low-income countries.

- Allow participants a say in shaping the Forum's political and practical directions for the WSSCC's future operations.
- Retain its emphasis on advocacy for the MDGs.
- Be structured around a strong thematic foundation.

Attendance at future fora

- Eighty-eight percent of respondents would attend a future Global WASH Forum.

What were the immediate impacts of the Global WASH Forum?

- Rather early to judge impact and difficult to attribute directly to the forum alone.
- However, there are some examples of government initiatives in India, South Africa and Uganda that demonstrate certain impacts. These include an increase in budgetary allocations for sanitation issues and the establishment of national WASH coalitions.

Did the Global WASH Forum have a wider impact?

The main impacts identified are the following:

- Sanitation is now recognized as an important issue at the national level.
- The Roadmap and its recommendations for the CSD 13 and the United Nations MDG progress review (September 2005) will result in a wider impact of the Forum.
- Attendance at the Forum had a catalytic effect resulting in enhanced activity at regional and local level.

What difference did attendance make?

Respondents benefited from attending the Global WASH Forum, owing to the following points.

- An opportunity for networking and establishing new contacts.
- The opportunity for new learning, knowledge exchange and information sharing.
- Increased motivation.

In terms of relevance, how is the Global WASH Forum distinct from other global events in the sector?

Key highlight: this was not well articulated by most respondents. Answers ranged from the Forum lays emphasis on practical results, to having 'wide representation', being 'truly global', and bringing an emphasis to sanitation.

What are the benefits to the individual of the Global WASH Forum?

- Networking and establishing new contacts.
- New learning, knowledge exchange and information sharing.
- Motivation for individuals.

Can the Forum continue to contribute to the needs of the sector?

Broadly, yes. Responses coalesced around the views:

- It is the only global event with an emphasis on sanitation.
- It provides a vital opportunity for networking on a global level.
- It indirectly applies pressure on the governments to act.

In terms of effectiveness, what outcomes have been delivered?

- There was a general lack of knowledge of what the Forum outcomes were supposed to be beyond the roadmap for CSD 13 and to some extent, the development of national WASH coalitions.
- The roadmap was cited by most respondents.
- It was stated that national coalitions have been established in some countries.

Political presence at the Global WASH Forum

- The consensus was that the political presence at the Forum was very important, very significant and should be encouraged at all global events in the sector.

Organization of the Global WASH Forum

With regard to Forum logistics, respondents felt the following:

- The venue was too lavish. This raised the question about whether the financial resources used in holding the Forum could not have been better used in improving access to sanitation.
- The number of participants in the Forum events was too large.
- A number of stakeholder groups were missing from the Forum, notably women and the younger generation (future leaders).
- Accommodation arrangements fell short of expectations of a global event. There was significant dissatisfaction.
- There was inadequate provision of translation services, especially given that the Forum was held in a Francophone country and the Forum language was mainly English.
- There were severe shortcomings in transport arrangements.
- Ineffective timetabling of events and housekeeping information impeded the Forum experience.

Content of the Global WASH Forum

- The thematic approach was excellent.
- It could have had better advocacy.

Structure of the Global WASH Forum

Respondents felt the following:

- Ideally, the Forum should take the lead in articulating priorities for the sector.
- The Forum should have focused on networking, learning and advocacy.

- The branding and identity of the Forum as a WSSCC event did not come through strongly enough. Some respondents did not think the Forum was in any way related to the WSSCC.

Parallel events (learning events, partner events and exhibitions)

Respondents felt the following:

- There were too many parallel events. A reduced number of carefully targeted events would have been more effective.
- There were too many events of interest to specific stakeholder groups that overlapped. This diminished the potential benefits of these events.

Keynote speeches

Respondents felt the following:

- The keynote speeches occupied too much time.
- The accent was on rhetoric. The general view is that they were ineffectual and a missed opportunity.

Thematic sessions

Respondents felt that the thematic approach was very good and that the choice of themes was relevant. However:

- There could have been fewer sessions, perhaps allowing more time for discussions.
- Some important themes and topics were excluded.

Conclusions and recommendations

The Evaluation team reached the following conclusions about the 6th Global WASH Forum.

A. In terms of sustainability

- The Forum has won endorsement from participants and the overwhelmingly sentiment was that it should continue.
- It is judged to still have a contribution to make in meeting sector needs.
- The purpose of any future WASH forum needs to be made much clearer and the added value of the forum needs to be better articulated to the participants.
- Respondents from the Global North were more inclined to regard the Forum as a conference. Those from the Global South, though not necessarily endorsing it as an approach, were happy to accommodate what was perceived as the "United Nations-type process" of the Forum. This lack of clarity is perhaps more to do with the change in the focus of the WSSCC away from a structure based around working groups to creating a focus on advocacy, a change which it would seem many people were unaware of, or had difficulty in accepting.

Recommendations

- The Global Forum should continue to be held, although the frequency with which it occurs should be reduced.
- The necessary emphasis on advocacy should be maintained and strengthened, although the Forum should also keep a thematic foundation.

B. In terms of impact
Whilst it is too early to make clear judgements on impact:

- The Forum has contributed to maintaining the global profile of sanitation.
- It has inspired and motivated a number of participants to action.

Recommendations

- The focus on WASH should remain.
- Opportunity (time and space) should be built into the programme of subsequent fora to ensure that participants can network effectively and establish new contacts and partnerships, as well as attending the main and parallel sessions.

C. In terms of relevance

- The Forum provides networking and knowledge-sharing benefits to individuals and broadly contributes to the needs of the sector.
- The distinctive attributes of the Forum are not sufficiently well articulated.

Recommendations

- The distinctive nature of the global fora needs to be demonstrated and communicated with greater clarity to the participants.

D. In terms of effectiveness

- The Forum followed a good and useful thematic approach which was well received.
- The content of the Forum was also considered to have been appropriate.
- The political presence at the Forum was important.
- The main promised outcomes have been delivered and, significantly, the Dakar Roadmap has been endorsed by the Intergovernmental Preparatory Meeting of the CSD 13.
- The Forum suffered in organizational terms because of insufficient in-house capacity to organize an event, its size and complexity, especially given the large number of side events and the political dimensions.
- A negative consequence of this was that the potential impact of the thematic structure was diluted.

Recommendations

- Consider different options for organizing and managing the Forum, e.g. internally managed by secretariat, externally managed by conference contractor; combination of internal/external based on specific functions.
- The number of main and parallel sessions should be reduced.

E. North–South divide

- We observed a strong North–South divide amongst respondents. This delineated the most extreme differences in responses within the overall respondent group. Those from the Global North are much less likely to come to a future Forum, have greater concerns with the nature of the venue and have less concern with political presence than do those from the Global South.

6.10 Relevance of the Analysis for a Megaconference

In spite of the difference in size and content, the issues raised are somewhat similar for the case of a megaconference. It is, however, interesting to note that a significantly smaller conference has been able to influence the global policy development in a more significant way than the huge megaconferences. The impacts and the outputs are identifiable and can be judged, though not with considerable accuracy but to a greater extent in terms of the follow-up in the immediate future. The above analysis shows that there are a number of areas that need to be improved for a smaller conference and possibly can be done through such independent evaluation. However, it is almost impossible to judge from the megaconferences excepting a general feeling of those who participated in it. Generally, it is now widely accepted that no single water megaconference, except the Mar del Plata, has significantly influenced any major policy or can claim to develop a clear theme and provided a solution.

Though there are many clear impacts and positive response, there are many loop holes and defects even for a smaller conference like the Global WASH Forum. Being the principal organizer, I recall how I debated against hosting this conference owing to the logistics and cost and tried to make as much out of an inherited situation. However, the silver lining of the conference was the predominant presence of the people from the ground with experience and they clearly responded to the question. They wanted the Forum to be held because it is the only one of its type with sanitation at the centre, and clearly with people in the focus. At least, the Dakar WASH Forum had put people in the centre, and through Dakar, to the United Nations itself. The process has started and the real impact can be seen through the progress of the issues at country level. This is a lesson that the organizers of future megaconferences should consider.

6.11 Conclusions and Recommendations

It is difficult to say when, where and whether the water circus will ever stop. However, the overdoing of these megaconferences has been counterproductive to the cause of advocacy and to the cause of water and development. It is difficult to see what impacts the Mexico Forum will bring, being held so close to the CSD13 and the United Nations General Assembly in September 2005. Instead, if it was held with the same support and enthusiasm after 2 or 3 years with a clear focus on the review of the progress of the CSD recommendations, it could have served as a global platform for measuring progress outside the United Nations System. The same will be true for any future forum, as without a clear political and implementation strategy they will always remain as semi-technical conferences, without any sharp focus and purpose. To find a solution out of this imbroglio the following is suggested for consideration.

- There should be a moratorium on the Forum for at least the next 5 years.
- Only two major forums on water should be organized between now and 2015, say at the end of 2010 and 2015, to review the progress on the global scenario on water and sanitation and efficient and equitable water management.
- The annual Stockholm Water Week can be used for serious policy development platform, in between these global forums.
- More emphasis should be on local and national level actions. Accordingly, prior process of national and sub-regional consultations, based on practical issues, is important.
- Smaller conferences on sanitation and hygiene which are often neglected at such global water forums should help the advocacy at regional and local levels.
- Focused forums or conferences, with clear outputs, linked with formal process of the United Nations, may prove to be useful in terms of their effects and impacts.
- Even for a smaller conference, communication between the organizer and the target audience is important and often not established. The situation is significantly worse for the global megaconferences.
- Every conference must revisit its pledges and outputs and assess the impacts regularly after the event is over for at least 5 years.
- Donor nations must use their funds for country-level actions and *not* for such forums.

6.12 Quo Vadis?

The juggernaut of the megaconference will continue, as this has become an industry. The only way to face it is to ignore it. However, is that a good design? Is this platform really a space for the people from the South, and South within the South? This does not seem to be the case. We should keep an open mind and look ahead for an answer. In the final analysis, water will always find its own level and so will the megaconferences.

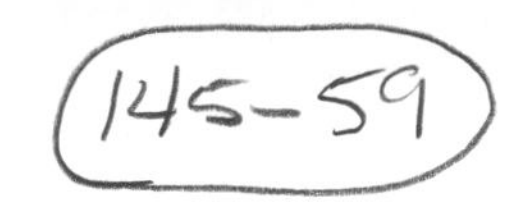

7 Evaluation of Global Megaconferences on Water

Asit K. Biswas and Cecilia Tortajada

7.1 Introduction

In the area of water, megaconferences are of comparatively recent origin. A megaconference in this context is defined as an event attracting more than 1,500 participants. Historically, one would be hard pressed to identify many water conferences that attracted more than 1,500 participants prior to 1975. The participants in this context are defined as people who participated in the presentations and discussions, and not someone who participated in ancillary events as visitors.

The definition of who is a participant is an important one because in recent years, there has been considerable "inflation" in the estimation of the number of participants. This is because, at least implicitly, one of the main indicators of success of a megaconference, if not the primary one, has been the number of participants it is supposed to have attracted, and not its outputs or impacts. Thus, the organizers often may include the number of visitors to ancillary events like exhibitions as participants, even though they had absolutely no role in the main meeting itself, and they may have visited the conference venue only once to see the exhibition or attend some minor events. For example, according to the World Water Council, the Japan Forum attracted 24,000 participants and the Mexico Forum had 20,000 participants. Even if all the seats available at these two Forums were taken, which they were not even during the opening plenary sessions, even less than half of these numbers could have been accommodated! For example, a bus-full of school children who may visit an exhibition, linked to a megaconference, can be considered to be visitors but they are certainly not participants to the discussions or what happens within the Forum itself. Thus, in the future, some clarity and consistency will be needed to collect the real statistics in terms of the number of participants for evaluation purposes. Because of this "flexibility" to define who is a participant, various estimates are now available as to how many people "participated" in any specific event.

7.2 Impacts of Megaconferences

None of the water-related global megaconferences has ever been evaluated independently, objectively and comprehensively. Their impacts and cost-effectiveness are now basically unknown, even though their costs have generally escalated very rapidly with each succeeding event. It should, however, be noted that very few megaconferences on any field have ever been objectively assessed in terms of its results and medium- to long-term impacts. Thus, water megaconferences are not an exception: it is simply part of an overall problem in this area.

In order to fill this gap, the Third World Centre for Water Management, with the support of the Sasakawa Peace Foundation of the United States conducted a global analysis on the perceived outputs and impacts of such events on the water sector, as well as their strengths, weaknesses and lessons learnt. The events specifically considered for this evaluation were Mar del Plata, Dublin, Rio and Johannesburg Conferences, Bonn Freshwater Conference and the first three World Water Forums.

A two-pronged approach was used for the evaluation of the impacts and the effectiveness of the megaconferences. The first part of the process was to invite a select group of experts who are well-versed with international water-related activities and are renowned for their acumen, objectivity and scholarly work, to review and assess the outputs and impacts of these events. One of the invited authors was also the Deputy Secretary-General of a major megaconference organized by the United Nations in the 1970s, who also happened to be a water expert. Vast majority of these authors had participated in one or more water-related megaconferences. In addition, all the authors were carefully selected and then specially invited to prepare assessments from their personal perspectives. They represented different countries, disciplines, sectors and institutions, both national and international, and also had diverse professional backgrounds and experience. The primary objective was to collect different objective views and analyses of the events. The authors of all these invited papers, as well as a group of additional selected water experts, were then invited to review and critique all these contributions at a special invitation-only workshop that was held at Bangkok, Thailand. In the light of these discussions, the authors then finalized their papers. These final papers are included in the present book.

The second component of the evaluation was a questionnaire survey of water professionals to find out their personal views on the outputs and impacts of these events. An open-ended questionnaire was sent to 2,698 people from all over the world, among whom were all the members of the World Water Council and International Water Resources Association, participants of selected major international and national water conferences, including Stockholm Water Symposium, and other professionals interested in water-related issues from academia, government, private sector and NGOs. The study also attempted to get a list of participants to the Japan Water Forum to solicit their views, but this request was declined. Questionnaires were also sent to participants who did not attend any of the megaconferences in order to see what impacts these events may have had on their institutions and/or on their work. Participants were specifically requested to give their own personal views and not of their institutions.

Special care was taken in order that the survey was universal and unbiased, and that study did not target any specific groups, countries or institutions, in order that as objective and comprehensive information as possible could be obtained. All the persons contacted were specifically informed that their responses would remain strictly confidential, and that these would not be shared with any other institution or individual not directly associated with the project under any circumstances. This explicit undertaking was given to ensure that all the respondents could freely give their own candid views, without any potential ramifications on their careers in the future. Accordingly, the access to the responses was strictly restricted only to the core study group who analysed the questionnaires.

In addition to this global questionnaire survey, some special countries or regions were selected for in-depth analyses. These special studies were focused on India, Bangladesh, Japan, Scandinavia and Southern Africa. Japan and Southern Africa were specifically selected since these countries/regions were host to two megaconferences: Japan for the Third World Water Forum and Southern Africa for the UN Conference on Sustainable Development. Scandinavia was selected since the Stockholm Water Symposium has become an important annual global event on water (in recent years, this Symposium has transformed itself to a megaconference), and also because many of the participants to the questionnaire survey came from Scandinavia. India and Bangladesh were selected because water is an important requirement for the social and economic development of both of these countries.

For each of these national/regional studies, internationally well-known and objective water experts from the region were invited to conduct the studies. For Japan, the questionnaire was translated into Japanese, and the survey was conducted in Japanese to overcome possible linguistic constraints. Each national/regional study leader sent out similar questionnaire like the one used for the global survey to the prospective correspondents in their study areas. In addition to the questions asked in the main questionnaire, the national/regional ones also had some additional questions that were considered to be relevant and important for the specific study areas. Thus, with the overall global study, and specific national/regional studies, the views of a very large number of people interested in water issues were canvassed and obtained. These regional analyses are included in this book.

During the analysis of the questionnaires, it was noted that personal views of some of the correspondents were very different to the views of their institutions. This was unquestionably due to the fact that the questionnaire very specifically requested personal views, whatever they may be. Also, the views expressed during the survey were assured to be totally confidential, and thus the respondents did not have to be politically correct, or worry about personal interrelationships with the sponsors and organizers of the events, or be concerned with potential backlashes, especially when their views were not complimentary on these events. This philosophy was also used for the Bangkok workshop, where the participants were very specifically invited in their personal capacities, and this factor was stressed in both the letters of invitation that was sent to them, and also during the workshop discussions. It was further stressed during the workshop that it was being organized under Chatham House rules in order that the participants could freely express their views, without any possible ramifications later. The present chapter is an analysis of the questionnaires that were returned.

7.3 Global Perceptions of Impacts

Altogether 2,698 questionnaires were sent electronically to the people interested in water from governmental, intergovernmental and international institutions, academia and research institutions, private sector and NGOs, both national and international. The total universe represented 121 countries. Even though the latest available addresses were used, 372 questionnaires bounced back.

Out of the 2,326 questionnaires which presumably reached their targets, 651 responses were received. This is a response rate of 28%. Considering the global, multi-institutional and multi-sectoral nature of the survey, and considering the possible sensitivities of the answers, the response rate can be considered to be quite reasonable. Of this number, 89 respondents said they not only did not attend any of the megaconferences, but also they knew nothing much about them in order to make any comment which could be considered meaningful in the context of the study. This itself is an interesting finding, since it indicated that some 14% of the respondents were not even aware of the proceedings and the results of the megaconferences, let alone of their possible impacts on the water sector. Of the balance of the 562 respondents, their personal participation rates in the various megaconferences are shown in Table 7.1.

Table 7.1 Percentage of respondents participating in megaconferences

Number of megaconferences attended by the respondents	Percentage
None	46.59
One	25.51
Two	11.36
Three	8.33
Four	4.54
Five	1.51
Six	0.37
Seven	0.37
Eight	0.37

7.4 Strengths of Megaconferences

Participants were asked to identify what in their views were the three most important strengths of the megaconferences. No list was provided of the perceived strengths from which they could select the most appropriate ones. Accordingly, the participants had to do their own thinking and then provide their own personal views. The main strengths identified by the correspondents the most were the following:

- Increasing awareness of water as a global concern, and of multidisciplinary, multi-sectoral and multi-institutional nature of the water problems and their possible solutions;
- Increasing awareness of the current water problems among various sectors of the population, including the general public and the media;
- Interactions of diverse views, opinions and visions under different settings;
- Better understanding and appreciation of how different countries have approached to solve somewhat similar water problems and what have been the results;
- Bringing together interested/relevant parties to one location for discussions of different issues from different perspectives and interests;
- Providing an opportunity for developing countries to raise their problems and concerns in an international setting so that the world as a whole realizes the complexities associated with their solutions, including the appropriate financial and institutional constraints faced;
- Identifying critical water problems and important global, regional and national water-related issues, which subsequently could become part of the professional agenda for additional discussions in other fora;
- Listening to different opinions on somewhat similar problems and issues, and be aware of the reasons of these differences;
- Ensuring increased attention on water issues as governments and institutions have to define and justify their positions during these events;
- Enhancing political support to domestic water agenda because of global support and recognition;
- Promoting professional–politician–NGO interactions;
- Focusing attention on specific water issues which are not receiving adequate attention at present;
- Meeting and listening to global water experts from different sectors, disciplines and nations to get more knowledge and better appreciation of water problems and their solutions from different parts of the world;
- Raising the profile of water-related issues in the national and the international media during the events;
- Meeting old friends, making new ones, and enhanced opportunities for networking with people from different parts of the world; and
- Exchanging ideas and information on technical, economic, political, social, environmental and legal situations and trends on water-related issues and problems, and their institutional implications.

7.5 Weaknesses of Megaconferences

Correspondents were similarly asked to identify three major weaknesses of the megaconferences. Again, no list was provided for this selection. Significantly more weaknesses were identified by the correspondents compared to the strengths.

The main weaknesses identified can be classified under the following categories. It should be noted that they are not in any order of priority.

- Too many issues are discussed superficially and often dogmatically, with too many poor-to-mediocre sessions which often repeat what has been said or written numerous times earlier;
- Too many presentations and set speeches which do not say anything new or interesting, and not enough time is available for proper discussions and interactions between the participants;
- Too many conflicting views in different sessions, and no real attempt to reconcile different views, or to assess them objectively, or strive for a consensus;
- Very poor (mostly non-existent) efforts to disseminate the results and background papers and documents, both before and after most of the events; unless one attends the events, or even a specific session, documentation is simply not available; documentation available after the events is far too general and superficial for any specific use; only exception has been the UN Water Conference at Mar del Plata for which detailed documentation is still readily available some three decades after;
- Overall planning and management of the events leave much to be desired; inadequate or inappropriate strategic thinking from the organizers as to what they wish to achieve from these events; interactions between participants are mostly superficial; sustained interactions are impossible because of the large number of participants; sessions and events are conspicuous by the absence of any discussion on long-term, or even medium-term, water-related trends and developments;
- Participants come primarily from the water sector and thus solutions are sought almost exclusively from within this sector; no attempt is made to consider multi-sectoral approaches, which is essential for solution of complex water-related problems;
- Conferences primarily deal with the sponsors' and the donors' agenda, who have limited knowledge, understanding and appreciation of water problems of developing countries;
- Events, outcomes and declarations, to a great extent, are controlled by a small group of individuals and institutions from developed countries, who are often the same from one megaconference to another; thus, superficially, it may appear that there is true stakeholders' participation and consultation during the planning process, but in reality the main agenda is controlled by a very few selected individuals and institutions; in other words, while on the surface there is a veneer of extensive stakeholders' participation, in reality the process and outcomes are controlled;
- Megaconferences have degenerated to be the likes of fairs or festivities, rather than being serious events; outcomes are mostly predetermined by certain groups, with preconceived ideas and hidden agendas rather than being generated from free, frank and true interactions between the participants during those meetings, as invariably claimed by its sponsors and diehard supporters;

- Too much repetition from one conference to another which often promotes bandwagon effects in many areas, and contributes to very little real progress in solving the actual water-related problems faced by different countries of the world;
- Unnecessary large nature and format make the events impersonal and forgettable experiences; numerous activities are redundant or peripheral and thus have no near-terms value, let alone over medium- to long-term; events often degenerate into ritualistic fanfare and self or institutional publicity, or restatements of the obvious;
- No attempt is made to prioritize critical water issues; thus wheat and chaff receive the same level of attention;
- Outputs lack specificity, cohesiveness and relevance, and often they are packaged in a new bottle but the wine continues to be old, even very old;
- No thought is given to their implementation potential, or who could implement them and also from where the funding for their implementation could come from;
- Policy dialogues are dominated by certain national and international institutions, which have very specific ideas, agendas and dogmas that they want to promote or perpetuate. These institutions are well-funded and, accordingly, can participate in all the preparatory meetings, which individuals or institutions from developing countries cannot. They primarily come to advance their agenda, power and visibility. Not surprisingly, their ideas mostly prevail, since they mostly control the process and the outputs, even when participants propose different but more efficient and implementable solutions;
- No political commitments for implementing declarations and commitments at national, regional and international levels;
- Raise very high expectations with high rhetoric, which are never fulfilled, as a result of which deep frustrations set in later;
- Seldom provide any new insights to the future global, regional and national water scenarios, except in somewhat general and superficial terms; discussions are invariably on the problems and solutions of the present or of the past and very seldom of the future;
- No attempt is ever made to objectively evaluate the performance, outputs and inputs of the events. In fact, the organizers would be hard pressed to define what exactly are the objectives of the events, and the types of outputs and impacts that should be reached in order that these meetings can be considered to be successful. The process has become more important than the end objectives; sponsors never discuss what in their views are the indicators of success of these meetings;
- Because of the global nature of the megaconferences, the linkages between them are mostly non-existent. There is thus no continuity, since they are primarily planned, designed and organized as discrete events;
- There are no mechanisms to promote and assess possible follow-up activities, nationally, regionally or internationally;

- Results are often lost opportunities with sanitized waffle with politically correct posturing, phraseology and adherence to, or promotion of, the prevailing bandwagons;
- By making a deliberate attempt to please every government, and declining to offend or ignore anyone, the Ministerial Declarations avoid hard choices in terms of priorities and specifics. They often degenerate to general and sanctimonious statements which are of very limited use in policy and planning terms to any country. In addition, these general statements have been made so many times before that no one takes these declarations seriously any more, including the ministers themselves;
- The world is very heterogeneous, with different physical, social, economic, environmental and institutional conditions. While the types of problems faced by many countries may be similar, the Forums basically discuss and promote monolithic solutions which implicitly assume that one size will fit all. Since this seldom is the case in the real world, a significant part of the discussions become irrelevant or inappropriate at least in terms of understanding of the problems and their possible realistic and implementable solutions.

7.6 Cost-Effectiveness

Overall, the respondents of the survey were very positive of the cost-effectiveness of the UN Water Conference at Mar del Plata. The overwhelming general consensus was that no other megaconference exceeded, let alone equalled, the impacts of the Mar del Plata. After Mar del Plata, the respondents felt that the Rio Conference also had discernable impacts on the water sector, both nationally and internationally, since it put environment firmly on the agenda of the water institutions. The respondents further believed that the Bonn Conference and the first three World Water Forums have not been cost-effective, and their impacts, if any, have been conspicuous by their invisibility.

Overall, the World Water Forums very especially came under severe criticisms from the correspondents for the following reasons:

- Poorly organized with the fundamental strategic error of organizing too many sessions, often on very similar topics, which means it is very difficult, if not impossible, for the participants to attend the sessions they want, and receive a consistent message, or get an overall picture of the results of the discussions on a specific issue;
- The main criteria for success appears to be how many people participated in the Forum, but not on the quality and relevance of the presentations, discussions and outputs. Very little thought is given during the planning process on how the proceedings could impact upon water planning and development processes and practices of different countries and international institutions;

- Costs for each succeeding Forum are going through the "ceiling to the sky", and thus it is necessary in the future to prune non-essential activities and events, and focus on result-oriented, doable, and useful activities;
- All "omelettes need eggs". Good ones are worth the efforts and the eggs. Bad ones are waste of eggs and efforts. Water Forums have become expensive events with very limited outputs like "bad omelettes";
- Cost-effectiveness can be increased somewhat, even with the existing arrangements, if serious efforts are made in terms of collecting, editing and disseminating good information that were presented, and generated during the events as well as through the processes leading to the events;
- Except for meeting old friends, and seeing new cities, the Forums now provide only limited benefits to most of the participants. In fact, six months after the event is over, one wonders what the fuss was all about, and what were the lasting results, if any. However, the events are useful to organizations like the World Water Council since it provides it with the main *raison d'être* of its existence, and considerable income from the events. Its other sponsors get some international visibility and power. However, the overall cost-effectiveness of these Forums for the world as a whole is very low.

Nearly 90% of the respondents felt that in the light of the experiences from the past Forums, a determined and comprehensive attempt should be made to redesign/restructure/rethink the way megaconferences are organized in the future in order that their impacts and cost-effectiveness can be increased very substantially.

7.7 Key Lessons

The general view that emerged from the survey was that the megaconferences have their own momentum, and they satisfy the needs and the agendas of certain specific institutions and individuals. Accordingly, they are likely to continue, at least for a while, in their present format, perhaps only with marginal and incremental changes, irrespective of what the majority of water professionals and water-related institutions may think about their relevance, impacts and effectiveness. Accordingly, it is somewhat unrealistic to expect that the next 2 to 3 water-related megaconferences will be materially different from the earlier ones in terms of process, structure, results or impacts. Especially for the World Water Forums, the perception is that the same group of institutions, and also many of the same individuals, that were responsible for organizing the earlier Forums are likely to continue to be the driving forces behind the arrangements for the future ones. Thus, at most, irrespective of the relevance and cost-effectiveness of the past Forums, one should realistically expect only minor changes in the foreseeable future.

The main lessons that could be learnt from the past megaconferences that were identified by the respondents are the following:

- Megaconferences generalize problems and solutions, even though the world is not homogenous. Equally, it is now well-known that one size does not fit all.

The global generalizations override a country's or even a region's specific needs and requirements, consideration of available management, technical and administrative expertise, and prevailing institutional and legal frameworks and financial capacities. The devil invariably is in the details and not in large generalized global talk-fests, where a good time can be had by most participants.

- It is not rewarding to assemble thousands of people with different views, agendas, interests and expertise, to discuss unreachable goals and targets, without any consideration of possible implementation of what often have been wishful-thinking conclusions, recommendations and declarations in the past.
- Megaconferences should be specifically focused on perceived needs and issues and they should have clearly stipulated goals and objectives. The process used for their organization should assure formulation of realistic, understandable and implementable recommendations for actions, and provide mechanisms to ensure the availability of adequate levels of funding to implement the recommendations. They should also bring to the attention of the participants latest scientific and technological developments, as well as what solutions have worked and what have not, along with the reasons for their successes and failures in different locations. Regular repetition of the same old water issues and problematics, as well as their so-called solutions, is a sure recipe for overkill in terms of impacts and relevance, both inside and outside the profession. These conferences appear to be reaching the point of diminishing returns. They do create temporary awareness of water-related issues, which evaporate quickly in the absence of follow-up actions, monitoring and evaluations. The events are thus rapidly losing their moral authority.
- Donors are still influencing the outcomes to suit their own views and agendas, irrespective of actual needs and requirements of developing countries. Many developing countries, irrespective of the rhetoric from both sides, continue to accept donor-driven agendas, priorities and solutions, which, not surprisingly, do not produce anticipated results. Progress thus continues to be marginal and suboptimal in the water field.
- It is necessary to consider country-specific or at most region-specific solutions and recommendations, which are implementable. The megaconferences should not be overloaded with pedestrian, outdated, irrelevant and dogmatic views and generalized presentations, most often this is the case at present. There are also no global solutions for a very heterogeneous world. The events should consider the importance of the co-existence of different paradigms, depending upon site-specific considerations.
- One way to look at the past megaconferences is that they are social events, which have been transformed into a form of water-related tourism. In fact, one will be hard-pressed to find even one-third of the total number of participants in most days at the formal sessions. One can thus legitimately wonder where are the rest of the two-thirds of the participants. These events often provide cover and legitimacy to the national, regional and international water institutions to do what they have done in the past, or are doing at present, or planning to do in the future, and then wrap around them the sessions, discussions and declarations emanating from these events to justify their past, present and future actions and

programmes to their superiors or governing bodies. The process gives a veneer of legitimacy and impression that the participating institutions are *au courrent* with the latest international ideas and cutting-edge developments as outlined during the discussions and as enunciated by the general recommendations and declarations. The events thus often degenerate to a self-serving process, rather than creating added values for countries, water profession or the world as a whole.

There is a general perception, especially among the organizers of these megaconferences that momentous advances and decisions are being made during these events, and these developments shape and chart the future of the water management in the world. Nothing could be further from the truth.

A retrospective analysis of impacts of these meetings indicates the following:

- These events have generally not only failed to give future directions for water management but also have mostly generated too few decisions that are realistic and can be implemented promptly and cost-effectively in different countries.
- They seldom provide a forum for adequate, objective and comprehensive analyses of current water problems, or sustained discussions and reflections of emerging trends which are likely to affect water management in the future. Irrespective of the continuous rhetoric of "business as usual is no longer a solution", the actual discussions indicate that the implicit assumption continues to be that basically "no business unusual" solutions are available. As a result, progress in water management due to these megaconferences can at best be marginal and incremental. The overwhelming focuses of discussions are often of the "SOS" (same old stuff) type, which one has heard time and again, sometimes even for over half a century!
- The processes used for the organization of these events, and subsequently much of the discussions at these meetings are often based on a cacophony of vested, entrenched and competing interests of national and international institutions, whose objectives often are to get some form of "blessing" or approval so that they can continue with their activities as before. Institutions like to be seen at these events, and try to publicize their activities and results, irrespective of their quality, relevance and implementation potential. Thereafter, they proudly proclaim their presence at these events, irrespective of whether they achieved anything substantial by their presence. There is often some form of paralysis, at least in terms of reaching consensus in many complex or controversial areas, because of the competing and conflicting interests and agendas of different institutions, both national and international.
- Many of the principal institutions that are directly associated with the organization of these events are gradually becoming ineffective because of their somewhat static view of water management in a rapidly changing and very heterogeneous world. Equally, many of these institutions are becoming increasingly ineffective because they were given responsibilities, often decades ago, when the global conditions were very different, or because they have unilaterally assumed additional responsibilities that far exceed their original authorities, as

well as their intellectual, technical and management capacities, and available financial resources.

- There is no question that during these megaconferences many interinstitutional or interpersonal deals get discussed, or even completed. However, the main objectives of these events should not include transactions of institutional or personal deals, for which better and more focused opportunities may exist elsewhere, and where such discussions can be conducted under more conducive and congenial conditions.
- A major question that needs to be asked and answered is why some 20,000 individuals, or even more, have participated in the third and fourth World Water Forums, if their impacts are mostly not discernable even six months after they ended. Probably the answer lies in the fact that these are basically social events, masquerading as serious meetings, with lots of receptions, good food and drinks, meeting many old friends and making a few new ones, and all under agreeable environments in interesting cities, and paid for by someone else. They can hardly be considered as a milestone or an important event in the history of water development, except by the organizing institutions. Several respondents to the survey felt that these megaconferences have basically now become "Woodstock of water", except Woodstock was a once in a lifetime event, but these water-related megaconferences are now being organized annually in one place or another.

7.8 Overall Impacts

Based on the survey conducted, nearly 44% of the respondents felt that megaconferences had no perceptible impacts, or at best marginal impacts, on them, or on their institutions, governmental, academic or private. Another 11.5% were even more negative on their assessments.

In contrast, only 7% felt that the events were "excellent," and 26% felt that the conferences have perceptibly changed the policies, programmes and projects of their institutions. It should, however, be noted that the majority of those who felt that the policies of their institutions have changed, referred very specifically to the UN Water Conference and the UN Conference on Environment and Development at Rio. The Rio event appeared to have injected strong environmental components in the policies and programmes of the various national and international water institutions. If Mar del Plata and Rio megaconferences are not included, the rest, according to the respondents, do not appear to have produced similar impacts. In fact, if the impacts of the Mar del Plata and the Rio are not included, the percentage of respondents who consider that these events had significant impacts come down to mid single digit. Another 11.5% of the respondents did not express any view. An important difference noted is that the respondents from US, Canada and Western Europe were noticeably more sceptical on the benefits and impacts of the megaconferences. In contrast, the participants from developing countries had a somewhat more positive view.

It should also be noted that several of the respondents felt that some of the impacts of the megaconferences have actually been negative since they have sometimes promoted inappropriate approaches and solutions that several institutions have later implemented with disappointing results.

7.9 Overall View on the Megaconferences

The overall views of the respondents on the megaconferences assessed are summarized in the Table 7.2.

Table 7.2 Views of the respondente on the megaconferences

Views	Percentage
The concept of such global conferences is good, but the current framework for organization needs to be changed radically. The events should be more focused and output-oriented. The main criteria of success should not be the number of people who attended the conference, nor the number of countries represented, but the quality of the discussions, results and their eventual impacts.	48.37
It would be desirable to organize regional meetings, dealing with regional problems, issues, solutions and institutions, and which could be focused and output- and impact-oriented.	30.70
The events have now become one big "water fair," with a lot of activities but without much thought being given to their relevance, appropriateness, outputs or impacts. There are no interlinkages between succeeding megaconferences, no clear focus, and their cost-effectiveness leaves much to be desired.	11.48
The global megaconferences in their current form are useful and cost-effective. We should continue with them, but only with marginal changes.	2.27
No view	7.18

The overall view of the respondents very clearly indicates that the people are now mostly sceptical of the benefits of the present form of the global megaconferences in the water sector. While they feel that there are needs for global and regional water policy dialogues, the general view is that these events should become more focused, and more problem- and solution-oriented, rather than being extended talk-fests. They should be planned with clearly identified goals and objectives in terms of achievements and impacts, which do not appear to be the case at present. Equally, there should be mechanisms in place to monitor the follow-up activities and impacts of the events, and also for objective evaluations by independent and capable experts of the processes used to organize each event, the event itself and its outputs and impacts. These evaluators must not only be independent, but also be perceived by the water profession to be independent, so that their assessments are considered to be accurate, objective, comprehensive and without hidden agendas or biases. The evaluators should not be linked to the conference organizers in any way to avoid pseudo-evaluations. Furthermore, these evaluations should not be kept confidential: rather, any one interested should have easy access to them. The evaluations should subsequently be used to improve the process, structure and other organizational aspects of the future megaconferences in order that their outputs are usable and their impacts become significant.

It should be noted that good and objective evaluations are only part of the story. Until and unless the organizers of these events seriously consider the results of the evaluations and then incorporate them appropriately during the planning process for the next event, these analyses are likely to be only paper exercises that will simply collect dust on various book shelves. Furthermore, and most unfortunately, currently there appears to be a tendency to shoot the messenger bringing any bad news, rather than consider the reasons behind such news, and then decide how the situation can be improved.

The respondents pointed out that for any large continuing series of events, there are invariably many vested interests, both personal and institutional, as well as inertia in terms of instituting any significant change, which collectively will favour mostly business-as-usual approaches with only incremental, but marginal, changes. It will require a courageous, enlightened and politically incorrect leadership to institute the necessary changes that are now clearly needed. Unfortunately, such enlightened leaderships generally are mostly in short supply in nearly all fields, including water. Without such a determined effort, the megaconferences are most likely to continue to attract large number of participants as "water fairs", but equally they will continue to have limited impacts on water management processes and practices, globally, regionally or nationally.

7.10 Alternatives to Megaconferences

During the Bangkok workshop, it was noted that the megaconferences are in fact only one small part of an overall global process, with numerous other actors all over the world, who are contributing to an exponential increase in understanding of better ways of planning and managing water. Vast majority of these actors have never participated in these events and never will. They are unlikely to do so in the future. Even if it is assumed that these events are attracting around 20,000 real participants, which is highly unlikely, these numbers, when viewed from a global perspective, are not that significant, especially when compared to the number of people interested in water issues all over the world. Viewed in another way, even this high number of 20,000 represents less than 1% of the people interested in water-related issues in a single major country like China or India. Thus, vast majority of people interested in water do not attend these meetings, or are even aware of them. The organizers of these events may have self-interest in conveying the impression and/or encouraging the perception that these global megaconferences, with numerous ministers and water experts from different countries of the world, are the places where seminal and momentous decisions on overall water management are made. Unfortunately, this is not the case. While these meetings may discuss most water-related issues under the sun, they are generally producing very few concrete and implementable ideas and decisions.

Among a number of barriers to effective actions to improve the effectiveness of megaconferences are inertia, with its related partners and attributes such as ignorance, scepticism, fatalism, etc., which are powerful forces that are likely to prefer the maintenance of the *status quo*, aversion to risk-taking in the face of many uncertainties and their implications, and vested interests in maintaining the present situation. This is particularly relevant for those who fear they may have to bear the lion's share of the financial and other intangible costs of any change in institutional and/or personal terms but without assurance of commensurate benefits or pay-off. Furthermore, and most importantly, any serious change will require strong and determined national and international leaderships. Equally, all these events will have to invariably wrestle with cacophony of entrenched competing institutional and personal interests which mostly will require extensive compromises to arrive at any consensus decision on any complex issue. The current process mostly results in the acceptance of the lowest common denominator types of actions, which are acceptable to most actors. They are highly unlikely to be at business unusual type of decisions that the world currently needs and clamouring for.

Because of numerous prevailing uncertainties, it is too early to predict the future shape of the global water megaconferences with any degree of confidence. However, what is certain that the current overwhelming global perception is that these megaconferences are not delivering the results that were anticipated, and business as usual is no longer an acceptable alternative. Whether there will be an emphatically courageous call from the water profession as a whole for changes in the structure of the megaconferences, in stark contrast to what has happened in recent years, only the future can tell. What is evident from the global survey is that there will be renewed and persistent calls for more changes which will ensure that business as usual is no longer an alternative for the way the past water-related megaconferences have been organized. What no one can foretell at present, with any degree of confidence, is how long it will take before the needed changes will actually occur.

PART II: COUNTRY-SPECIFIC ANALYSIS

8 Megaconferences: Some Views from Japan

Mikiyasu Nakayama and Kumi Furuyashiki

8.1 Introduction

As part of the world-wide survey on the impacts of water-related megaconferences initiated by the Third World Centre for Water Management in Mexico, the Institute of Environmental Studies of the Graduate School of Frontier Science, the University of Tokyo, was commissioned to conduct a country survey in Japan. The global survey was carried out in an attempt to assess whether a number of global conferences on (exclusively or inclusively) water issues, held in a grandiose scale almost incessantly in the past few decades all around the world, have generated any impacts on the water sector. And if they have, what those impacts are, whether positive or negative, as perceived especially among those working in water-related fields. The Japan country survey was conducted as part of such an attempt, in the form of questionnaire survey as suggested by the Third World Centre for Water Management. This report summarizes the findings of the country survey.

Although the authors believe that the views expressed from within Japan through this survey do reflect some reality, as discussed later in the chapter it is not quite appropriate, or it could even be dangerous, to try to come up with a solid conclusion regarding the effectiveness of megaconferences based on this single survey only. Therefore, it is the authors' hope that the findings from this survey will contribute to a discussion on the future course of similar megaconferences, which we assume will keep being held in this era of globalization. This analysis, therefore, is to provide some data for the discussion, but does not resort to an in-depth analysis of the survey findings to extract any specific conclusion. The authors hope that this global survey will also lead to a discussion on the evaluation methodology of the effects of such conferences itself.

8.2 Survey Method

The Japan survey team at the University of Tokyo prepared a questionnaire format based on the original one provided by the Third World Centre for Water Management. All the questions were translated into Japanese, and the format was slightly modified to ease the answering as well as summarizing processes. A few questions were added as and where the authors felt appropriate. Each modification and addition is mentioned in the following text. Care was taken, however, not to make the content of the questionnaire greatly differ from the original one, so that the clear comparison with the findings from other regional/country surveys may be possible at a later stage.

In accordance with the original questionnaire, the following conferences were the object of the evaluation in the Japanese survey:

1. United Nations Water Conference, Mar del Plata, 1977.
2. International Conference on Water and the Environment, Dublin, 1992.
3. United Nations Conference on Environment and Development (UNCED), Rio de Janeiro, 1992.
4. First World Water Forum (WWF1), Marrakech, 1997.
5. Second World Water Forum (WWF2), the Hague, 2000.
6. International Conference on Freshwater, Bonn, 2001.
7. United Nations Conference on Sustainable Development (UNCSD), Johannesburg, 2002.
8. Third World Water Forum (WWF3), Kyoto/Osaka/Shiga, 2003.

The questionnaire was sent by email to approximately 430 Japanese individuals, either working in the water sector or known to have an interest in water issues even if not professionally involved in the water sector. Those individuals were selected from the professional acquaintances of the authors, participants lists of past water-related academic meetings (not megaconferences) held in Japan or nearby Asian countries, as well as the list of session organizers of the Third World Water Forum available on the web. Forty-five responses were received (with a response rate of approximately 10%).

In addition, 12 responses to the original English questionnaire from Japanese experts were provided by the Third World Centre for Water Management to be added to the summary of the Japan survey. Although the slight difference in the questionnaire format made it difficult to perfectly merge these responses with the 45 responses received in Japanese, efforts were made to incorporate them in the following summary as much as possible. Therefore, this report can be considered as a summary of 57 responses in total.

Following suit of the original questionnaire, in the Japanese version the recipients were requested to express their personal views rather than try to represent the view of their organizations. It was also stated in the questionnaire that unless specifically requested, all the answers would be quoted anonymously. Therefore no personal name or affiliation of the respondents is mentioned in the following sections. However, just for information the 45 respondents of the Japanese survey include staff members of research institutions, government agencies (7 each), academic institutions (5), donor agencies, development consultant companies, non-profit organizations (NPOs), graduate students specialized in international water issues (4 each), international organizations (3), water-related industry associations and foundations (2 each), and a private company. Although the number of responses obtained was rather small, it can be said that respondents were of diverse backgrounds coming from (though not considered as representing) different interest groups.

The results of the questionnaire are presented in the following section one by one.

8.3 Results

8.3.1 Question 1: Participation

In the original questionnaire, this question was to ask whether the respondent has ever participated in any of the eight conferences targeted in this survey. In the Japanese version, it was also asked whether the respondent was aware of the fact that the conferences had been held at all. The authors felt that many Japanese, even those professionally engaged in a water-related field, are not necessarily familiar with some of those conferences, and it might be useful to clarify the general (perceptual) familiarity with different conferences. The result is as shown in Figs. 8.1 and 8.2.

In terms of participation, more than half (39 respondents, 68.4%) of the respondents participated in WWF3. Even discounting the fact that some of the questionnaire recipients were selected from the list of WWF3 session organizers to start with, this large percentage of participation in WWF3 is understandable considering the acceptability and wide publicity within Japan. Zero to only a few (1–3 respondents) had participated in the conferences before 2001, and five respondents (8.8%) were in the position to attend the UNCSD in Johannesburg.

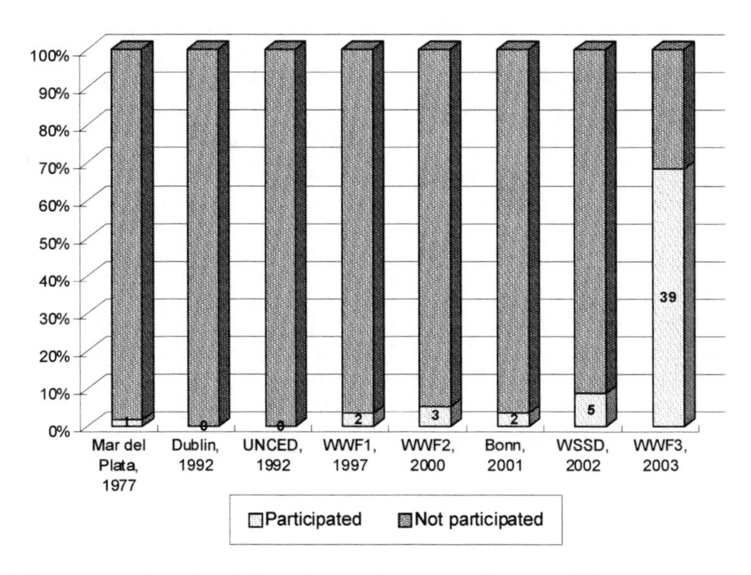

Fig. 8.1 Answers to Question 1: Participation in megaconferences (57 responses)

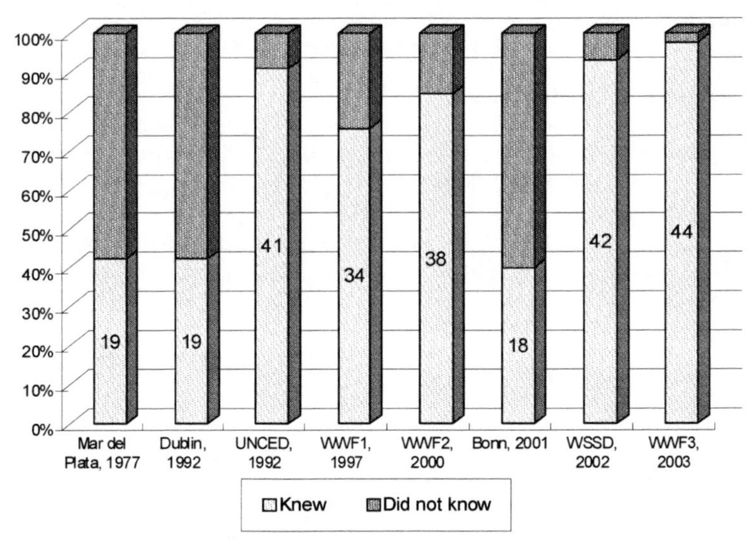

Fig. 8.2 Answers to Question 2: Awareness about the megaconferences (45 responses)

In terms of the awareness of those meetings, naturally WWF3 was known by almost all of the respondents for the same reasons as above, and WWF1 and WWF2 were also known by a large percentage of the respondents, possibly and partly in connection with WWF3. Over 90% of the respondents were also aware of the two conferences on broader environment and development issues and not only on water, UNCED and UNCSD, though they were held far away from Japan. On the other hand, only 40–42% of the respondents were aware of the other two conferences which focused exclusively on water (Dublin and Bonn). In other words, more than half of them were not even aware of the fact that these conferences had been held.

The nature of the conferences and the extent of media coverage within the country may have contributed to such an awareness gap. Whatever the reason behind this might be, as seen in the results of other questions, most of the following answers by the Japanese respondents seem to be based on their knowledge, experience and impression of WWF3, and to a lesser extent UNCED and UNCSD.

8.3.2 Question 2: Overall Views on each Megaconference

The original questionnaire of the Third World Centre for Water Management asked the respondents to grade each of the megaconferences in a scale of 0 (very poor) to 5 (absolutely excellent), setting 3 for 'average'. In the Japanese version, the scale was given from 0 to 6 with 3 for 'average', to make the number of positive and negative grades equal (0 = very poor, 1 = poor, 2 = relatively poor, and 4 = relatively good, 5 = good, 6 = very good). Also, the Japanese questionnaire specifically gave the option to write 'X' (N/A) when the respondent was not able to grade a conference.

The answers of the Japanese questionnaire and 12 answers of the original questionnaire are shown separately in Tables 8.1 and 8.2.

Table 8.1 Overall grading – Japanese questionnaire (45 respondents)

	0	**1**	**2**	**3**	**4**	**5**	**6**	**N/A**
Mar del Plata, 1977	0	0	0	5	2	1	1	35
%	–	–	–	11.1	4.4	2.2	2.2	**77.8**
Dublin, 1992	0	0	0	4	2	4	1	31
%	–	–	–	8.9	4.4	8.9	2.2	**68.9**
UNCED, 1992	0	1	1	4	9	15	3	12
%	–	2.2	2.2	8.9	20.0	33.3	6.7	26.7
WWF1, 1997	0	0	3	10	4	4	0	23
%	–	–	6.7	22.2	8.9	8.9	–	**51.1**
WWF2, 2000	0	0	2	11	7	5	0	19
%	–	–	4.4	24.4	15.6	11.1	–	42.2
Bonn, 2001	0	0	0	8	4	2	0	30
%	–	–	–	17.8	8.9	4.4	–	**66.7**
UNCSD, 2002	1	0	4	8	12	9	0	10
%	2.2	–	8.9	17.8	26.7	20.0	–	22.2
WWF3, 2003	0	2	2	6	15	13	3	3
%	–	4.4	4.4	13.3	33.3	28.9	6.7	6.7

N.B. Two larger percentages of each conference grading are highlighted in *grey*. The column for the average grading "3" is enclosed in *bold lines* to indicate the middle point. In the last column, percentages over 50% are highlighted in *bold*.

Among the respondents of the Japanese questionnaire, more than half were not able to grade the Mar del Plata Conference, Dublin Conference, WWF1 and Bonn Conference. Of those who graded these four conferences, none gave negative (below average) grades to the Dublin and Bonn conferences, while for Mar del Plata and WWF1 also most of those who were able to grade indicated the grade of more than average.

For UNCED and WWF3, more than half of the respondents gave a grade of 4 or 5 (53.3% for UNCED and 62.2% for WWF3). For UNCSD also, nearly half (46.7%) of the respondents graded as 4 or 5. If grade 3 (17.8%) is also counted, more than half of the respondents graded the Johannesburg Conference positively (more than average). For WWF2, many (42.2%) were not able to grade, but among those graded the largest portion gave 3, followed by 4 and 5. Here also more than half of the respondents (51.1%) gave positive (more than average) grades.

Table 8.2 Overall grading – Original English questionnaire (12 Japanese respondents)

	0	**1**	**2**	**3**	**4**	**5**	**N/A**
Mar del Plata, 1977	0	0	1	0	3	3	5
%	–	–	8.3	–	25.0	25.0	41.7
Dublin, 1992	0	0	1	0	2	3	6
%	–	–	8.3	–	16.7	25.0	**50.0**
UNCED, 1992	0	0	0	3	2	2	5
%	–	–	–	25.0	16.7	16.7	41.7
WWF1, 1997	0	0	0	3	3	0	6
%	–	–	–	25.0	25.0	–	**50.0**
WWF2, 2000	0	0	0	3	3	0	6
%	–	–	–	25.0	25.0	–	**50.0**
Bonn, 2001	0	0	1	2	0	0	9
%	–	–	8.3	16.7	–	–	**75.0**
UNCSD, 2002	0	0	2	1	2	0	6
%	–	–	16.7	8.3	16.7	–	**50.0**
WWF3, 2003	0	0	1	5	4	1	1
%	–	–	8.3	41.7	33.3	8.3	8.3

N.B. Two larger percentages of each conference grading are highlighted in *grey*. The column for the average grading "3" is enclosed in *bold lines* to indicate the middle point. In the last column, percentages over 50% are highlighted in *bold*.

Half or nearly half of the 12 respondents of the original English questionnaire did not grade the megaconferences except for WWF3. Among those who graded WWF3, nine people (75%) gave 3 or 4, evaluating it as "more than average". Among the rest of the conferences, Mar del Plata, Dublin, UNCED, WWF1 and WWF2 were graded mostly as positive or more than average. The Bonn Conference and UNCSD got a slightly negative grade of 2. None of the above conferences received the lowest two grades, 0 and 1.

Although the number of the respondents whose answers are compiled above is extremely small, it can be seen from the results that most of the Japanese respondents seem to regard the conferences in question positively or at least as "average".

8.3.3 Question 3: Impacts of Megaconferences

Question 3 in the original questionnaire read as follows:

Irrespective of whether you participated or not in these megaconferences, please select which one of the following comments most closely reflects your overall views on all the megaconferences as a whole:

A: The conferences were excellent. They have radically increased my knowledge-base, and have improved significantly my working practices.

B: The conferences have significantly changed the policies, programmes and projects of my institution. These changes would not have happened if these conferences had not taken place.

C: The conferences had at best a marginal impact on me and/or my institution.

D: The conferences had no perceptible impact on me and/or my institution.

E: It was pleasant to attend the conference(s), meet old friends and make new ones, but the conferences really had no lasting or visible impacts on me or my institution.

F: These were mostly forgettable events. For all practical purposes, it would not have mattered much whether these events had ever been held or not. They simply did not leave any footprints on water management.

In the Japanese version, the item G: "Other than the above. (Please specify.)" and H: "I do not know" were added. The result is shown in Fig. 8.3.

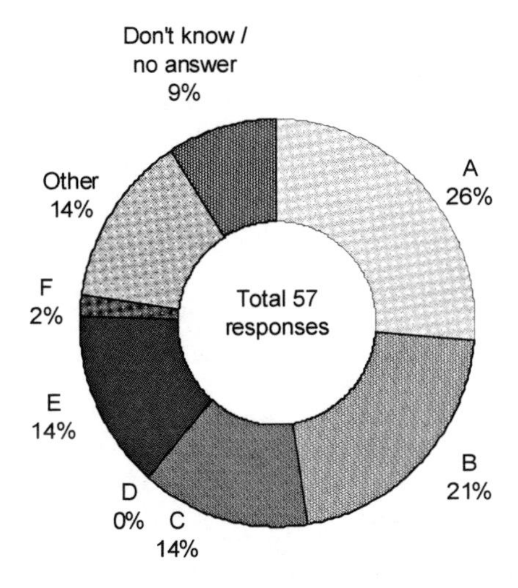

Fig. 8.3 Answers to Question 3

A large portion (47%) of the respondents gave positive answers as A or B. None gave the answer D (no perceptible impacts on oneself and/or his/her institution.)

Among the eight respondents (14%) who answered G: "Other than the above", two respondents said that the conference(s) had some positive impacts on themselves, but little impact on their institutions.

Another "Other" respondent commented that the first conference of any theme (including water) seems to generate big impacts in many ways, but from the second conference of the same or similar theme onwards, the scale of the meetings seems to get bigger and bigger while the discussion themes become more appropriate for smaller expert meetings.

Also, there was a comment that those megaconferences are important as a forum to understand the interests of different countries, and also to symbolize an agreement among them. Another respondent stated that those conferences are useful to learn what sort of projects would receive more money from the government.

The rest of the comments expressed by those who answered as G: "Other than the above" were negative ones stating that these conferences had not had any significant impacts (could possibly be counted as D).

The answers to this question show that the views on the overall impacts of those conferences were quite mixed among the Japanese respondents, with both the positive and negative views existing in a large portion.

8.3.4 Question 4: Strengths of Megaconferences

The question posed here was as follows:

> What in your view are the most important strengths of these megaconferences (list a maximum of three reasons).

The answers to this question are as listed in Table 8.3 in the descending order of the number of respondents.

Table 8.3 Answers to Question 4: Strengths of megaconferences

Strengths (advantages, merits)	No. out of 57 (%)
- Direct information exchange, opinion exchange	26 (45.6%)
- Awareness raising, information dissemination and formation of leading opinions in the society	21 (36.8%)
- Form common perceptions and understanding on the issues in question	14 (24.6%)
- Personal/professional networking (meet old friends, make new contacts)	12 (21.1%)
- Put together different interests of countries and make common decisions	7 (12.3%)
- Impact on domestic policies: by providing a basis or incentive for the improvement of domestic policies (whether it is realized or not)	6 (10.5%)
- Increase the significance of various (minority) stakeholders by providing them with opportunities to speak up	5 (8.8%)
- Enable the participants to feel the global trend on the issues in question (and to see the position, strength and weakness of one's own country objectively)	5 (8.8%)
- Provide efficiency in understanding, decision-making and actions	4 (7.0%)
- Enable global and political commitments to be made among the participating governments	4 (7.0%)
- Provide an opportunity to establish certain action target within a relevant institution based on the trend set by the conference	4 (7.0%)
- Provide an opportunity for developing countries to speak up and ask for more aid	2 (3.5%)
- International interactions for nurturing mutual understanding	2 (3.5%)
- Lead to promotion or better understanding of research activities	2 (3.5%)

There are a number of other answers provided by a single respondent each, including:

- useful to sustain the awareness on the issues in question;
- provide positive hope for international cooperation which in reality is a very difficult task;
- provide an indicator to evaluate the appropriateness of the following activities;
- provide a framework for discussion (whether it leads to any meaningful outcomes or not);
- able to cover overall water issues in local and global scales;
- able to collect big money by involving United Nations and governments;
- able to mobilize a large number of people;
- provide good business chances for private companies, (especially those from the West); and,
- the attention to one's own activities received at the conference encourage future activities.

8.3.5 Question 5. Weaknesses of Megaconferences

The question posed here was as below:

> What in your view are the most important weaknesses of these megaconferences (list a maximum of three reasons).

The answers to this question are listed in Table 8.4 in the descending order of the number of respondents:

Table 8.4 Answers to Question 5: Weakness of megaconferences

Weaknesses (disadvantages, demerits)	No. out of 57 (%)
- Lack focuses and follow-ups, ending up being too broad and general and difficult to see any concrete outcomes or impacts	22 (38.6%)
- Too expensive (low cost-effectiveness)	17 (29.8%)
- Too huge a scale to form a common understanding or to have substantial and realistic discussions, tendency to agree in general but oppose in details, limitation of time and space for detailed discussions	14 (24.6%)
- Dominated by a handful of people/entities (powerful states with strong interests, 'water specialists' of certain (mainly Western) countries, secretariat, those who are good at languages and negotiation skills, session leaders, big companies, etc.)	9 (15.8%)
- Too many participants and too many formal ceremonies or attractions, making the whole event a mere merrymaking	8 (14.0%)
- Difficult to understand what is happening for the general public (also due to the high participation fee and closed sessions), discussions and outcomes not communicated to the general public very well	5 (8.8%)
- Too much logistical burden	5 (8.8%)
- Lack of overall control	4 (7.0%)
- Repetitive	3 (5.3%)
- Difficult to catch up with what is happening and where, need to give up certain themes of interest when several sessions are held in parallel	3 (5.3%)

Table 8.4 (Continued)

Weaknesses (disadvantages, demerits)	No. out of 57 (%)
- Too expensive and not enough support for developing countries to participate	3 (5.3%)
- Participation itself becomes an objective, place for "professional travellers"	3 (5.3%)
- Difficult to establish a solid action plan for international cooperation in a real sense	2 (3.5%)
- Not academic or scientific (but rather political)	2 (3.5%)
- Tendency to put each individual water issue into compartments, and no correlation across different water issues	2 (3.5%)

There are a number of other answers provided by a single respondent each, including:

- difficult to come to a common agreement;
- difficult to sustain the interest after the conference;
- inflexible programmes;
- large gaps in research qualities of different countries revealed;
- difficult to evaluate the effect of the conference in figures (cost-benefit analysis);
- language barrier;
- gaps in follow-up actions in each country even after agreeing on a basic principle;
- recommendations or declarations which are difficult or unrealistic for developing countries are adopted;
- no legally binding outcomes;
- serves as a salon for high-level bureaucrats but a nuisance for those at the working level;
- minority opinions are disregarded; and
- "influence of mass psychology".

8.3.6 Question 6: Cost-Effectiveness of Megaconferences

The original question in the English questionnaire was as below:

Based on your perception of their outputs and impacts, what is your view of the cost-effectiveness of these events?

In the Japanese questionnaire, the format of this question was changed as follows:

Question: Which of the following comments is the closest to your view on the cost-effectiveness of these events?

A: All the conferences generated outcomes and impacts good enough to justify the cost.

B: Many of the conferences generated outcomes and impacts good enough to justify the cost, but a few of them did not.

C: Only a few of the conferences generated outcomes and impacts good enough to justify the cost, but many of them did not.

D: None of the conferences generated outcomes and impacts good enough to justify the cost.

E: Other than the above. (Please specify).

F: I don't know/No answer.

The result is shown in Fig. 8.4.

None of the respondents chose the answer A. Of the respondents, 44% (20) were of the opinion that only a few of the conferences were cost-effective and others were not, followed by 22% (10) who thought many of them were actually cost-effective.

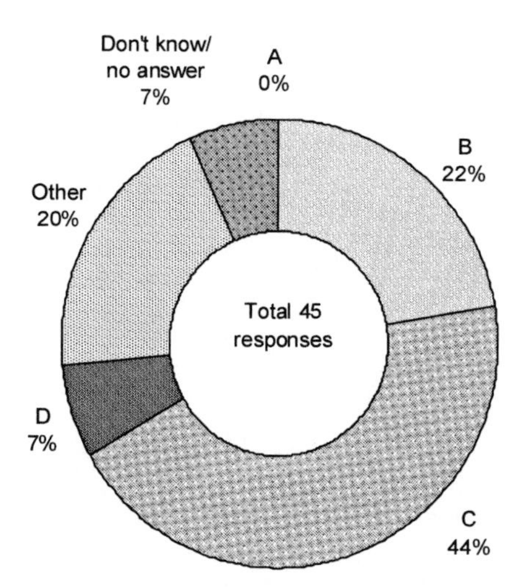

Fig. 8.4 Answers to Question 6

Among the nine respondents (20%) who answered as "Other than the above" and the 12 Japanese respondents of the original questionnaire, six respondents commented that the cost-effectiveness cannot be judged as there is no clear method for calculating it, though some of them also commented that the cost should be minimized as much as possible. Eight others expressed complaints and concerns regarding the wasteful use of money. Some of them also expressed their concerns on the conferences' diversion from the original purposes, which further lowers the cost-effectiveness. Two respondents accepted that a certain amount of waste was unavoidable for the smooth organization a conference, although "it is not desirable that conference fees also become higher accordingly".

Two respondents suggested ways to reduce the financial burden of the host country. One was to have participation of NGOs financed by their own governments rather than the conference's host country. Another was to organize such conferences in two phases:

> First, region (continent)-based conferences in which the organizers can expect more participants who cannot afford to travel a long distance. Second, a worldwide wrap-up conference with representatives from each region (continent). Region-based meetings may focus on more region-specific issues and clarify what they would like to bring to the global discussion table.

Only one respondent (to the original English version) stated that the conferences were cost-effective.

8.3.7 Question 7: Documentation and Information Dissemination

The original question read as follows:

Do you have adequate documentation (reports, papers, proceeding, etc.) from any of these conferences? If so, which ones? How useful has this documentation been? Overall, what are your views on the quality of the documents you have seen, and the information dissemination processes of these megaconferences?

In the Japanese survey, Question 7 was reformatted and made into Question 7a to 7d as below. Responses to each question will also follow.

Question 7a: Do you have adequate documentation (reports, papers, proceeding, etc.) from any of these conferences?

Result:
Yes: 40 (70.2%)
No: 11 (19.3%)
No answer: 6 (10.5%)

Question 7b: If you have answered 'Yes' in Question 7a, what sort of documents are they and from which conferences?

The result is shown in Table 8.5. Most of the documentation in the respondents' possession were naturally from WWF3. Most of the respondents did not specify what sort of documentation they had, but at least in terms of WWF3, materials mentioned included papers, proceedings, final report, programmes, handouts from participants, and the Ministerial Declaration.

Table 8.5 Answers to Question 7b

Documents from	No. out of 40 (%)
WWF3	35 (87.5%)
WWF2	4 (10%)
UNCED	2 (5%)
WWF1	2 (5%)
UNCSD	2 (5%)
Most of the conferences in question	2 (5%)
Bonn	1 (2.5%)
Dublin	1 (2.5%)
Do not know (owned by the institution)	1 (2.5%)

Question 7c: If you have answered 'Yes' in Question 7a, how useful has this documentation been?

The answers are listed in Table 8.6 in the descending order of the number of respondents.

Table 8.6 Answers to Question 7c

Use of documentation from megaconferences	No. out of 40 (%)
- Sometimes use as reference materials (for producing some work-related documents, research papers, teaching materials, for follow-up activities, or before attending meetings of a similar nature)	17 (42.5%)
- Just for information: to check the global trend and benchmarks on water issues, discussion themes and points	9 (22.5%)
- Rarely use	5 (12.5%)
- No use at all (One commented: "I would confess that most documents have been on bookshelves and made my shelves "nice-looking" or seem "professional".)	5 (12.5%)
- Not for use but it is kept because my (our) own report was published in it.	2 (5%)
- It is useful to some extent but disappointing considering the scale of the event.	1 (2.5%)
- Read while listening to session presentations.	1 (2.5%)

Question 7d: What are your views on the quality of documents you have seen, and the information dissemination processes of these megaconferences? Please choose from below:

7d(1): Quantity of documentation/information – Too much/Adequate/Too little/No idea

7d(2): Quality of documentation/information – Very good/Good/Average/Poor/Very poor/Cannot generalize as it varies depending on the occasion/Other/No idea

7d(3): Process of information dissemination – Good/Not good/Other/No idea

Answers are shown in Figs. 8.5, 8.6 and 8.7.

Comments of those who answered as "Other" in Questions 7d(2) and 7d(3) include (by one respondent each unless otherwise specified):

- Follow-up information as to how the things discussed at the conference are materialized (two respondents).
- A lot of general information on each country is given, but often detailed technical problems are not made clear in such information.
- It is enough to have the contact details of organizations that one can access in case any information is needed.
- Not enough information is given on the field reality.
- Not easy to understand. More concise information wanted.
- Information on CD-ROM rather than paper is preferable (two respondents).

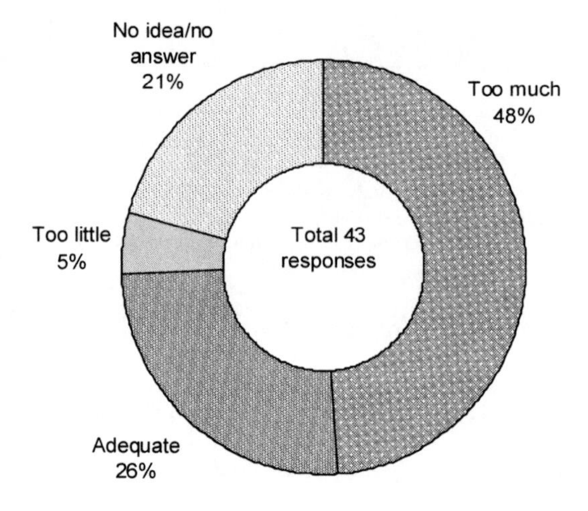

Fig. 8.5 Answers to Question 7d(1)

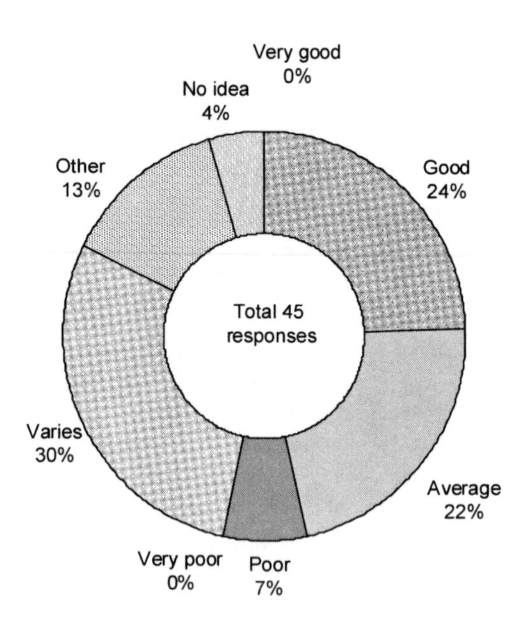

Fig. 8.6 Answers to Question 7d(2)

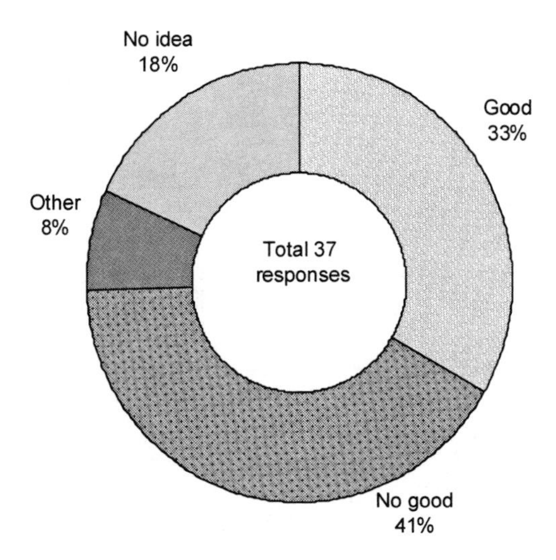

Fig. 8.7 Answers to Question 7d(3)

8.3.8 Question 8: Practical Results from the Megaconferences

This question was posed as follows:

> In your view, did any of these megaconferences have yielded positive, implementable and lasting results? If so, please give examples from the specific megaconferences at regional, national and/or global events.

Many of the respondents repeated the same answers as given in Question 4: Strengths of megaconferences. Namely, among those who did not specify particular conferences (20 respondents):

- awareness raising, nurturing common understanding among the general public and policy-makers (six respondents);
- impact on policies of the government, donor agencies (also on the amount of development aid in water sector) (three respondents):
- provide a framework for discussion, streamlining imminent global tasks on water (two respondents);
- personal/professional networking and interaction (two respondents); and
- forum for information and opinion exchange, comprehensive discussions (one respondent).

Three respondents (5.3%) expressed their regret that no practical results have been attained through the series of megaconferences.

Two respondents (3.5%) said there seem to be some practical outcomes but either she/he is not able to cite any specific cases due to lack of knowledge, or she/he feels that the mass media has not reflected the outcomes enough to prevail in the society.

One person felt that the results of the conferences targeted in this survey are not at the stage to be evaluated yet.

Among those who named specific conferences, six respondents (10.5% of the total of 57) felt that UNCED has yielded positive and lasting results in the form of impacts on environmental policies of the international community (United Nations and donor agencies) and the Japanese government. One respondent had a rather opposite view, doubting whether there has been any positive impact of UNCED on the current situation of the world.

Two respondents felt that UNCSD has generated some positive results, providing a policy guideline (though not legally binding) for United Nations agencies and confirming the action targets regarding the service of clean water and sanitation facilities.

A maximum number, 19 respondents (33.3%), commented on the results of the WWF3. Again, many of them repeated the same answer as given in Question 4:

- raised awareness, common understanding and perception on the water issues (six respondents). (though some of them added, "although no practical results have been achieved yet following WWF3");
- activated activities of citizen groups/NPOs (three respondents);
- opportunities for various stakeholders to speak up (two respondents).

The following 'results', also partly overlapping with some of the answers to Question 4, were perceived by a single respondent each:

- generated impacts on policies;
- useful personal network;
- agreed on some concrete actions;
- turned Japan's eye towards the outside;
- put together people of various professions concerned with water;
- able to show Japan's initiative;
- provided new impetus for the World Bank for water resources development and infrastructure building;
- possibly provide an incentive for Japan's overseas development aid to take certain water-related initiative.

8.3.9 Question 9: National Policy Changes due to Megaconferences

The question was asked as below:

> Are you aware of any policy change in your country which would not have occurred without one or more of these megaconferences? Is so, which policy or policies were changed because of these events?

Without specifying any conference, three respondents (5.3%) felt that those conferences have generated certain (limited) impacts on domestic laws and polities, as well as environmental guidelines of donor agencies (namely, Japan International Cooperation Agency (JICA) and Japan Bank for International Cooperation (JBIC)).

Also, one respondent commented that an increased understanding on water issues has led to certain increase in budget for sending water-related technologies overseas, while another person felt that the conferences have possibly increased the incentive for allocating more aid money for water-related development projects.

Two respondents stated that it was difficult to judge any direct impact on domestic policies, though they admit that the conferences might have contributed to raising awareness and understanding, and activated discussions.

Eight respondents (14.0%) said they have not seen any change in the domestic policies at all.

Among those who noted changes following specific conferences, one stated that the Dublin Conference had the greatest influence:

> In my view, the Dublin Principles except 3rd Principle of women's role in water supply, management and conservation, appear to have greatly affected Japan's water policy. More recently, the Millennium Development Goals by the United Nations Millennium Summit, though not mentioned in the current study, are having the greatest influence on Japan's water policy.

As many as 11 respondents (19.3%) felt that UNCED (as well as the concept of 'sustainable development' popularized following this conference) has brought some changes to Japan's domestic environmental policies, though not necessarily related to water. The examples mentioned included the establishment of the National Strategy for the Conservation and Sustainable Use of Biological Diversity (1995, following the ratification of the Convention on Biodiversity signed after UNCED), and the increasing emphasis on the 'environment' within the Basic Environment Law, River Law, and the Basic Agriculture Law. Three respondents commented that such policy shift towards the eco-friendly orientation has also led to the establishment of the Law on the Promotion of Nature Restoration (2002). Other specific policy changes mentioned by the respondents include increased support of the government for activities regarding global environmental issues such as technologies to reduce CO_2 emission and ozone-depleting substances, desertification, waste recycling, cleansing of polluted soil, etc., and movement towards the introduction of environmental taxes.

Five respondents (8.8%) mentioned WWF3 as a conference that has brought some changes to the domestic policies. One respondent mentioned the establishment of the River Environment Section within the Ministry of Land, Infrastructure and Transport (MLIT) as an example of 'policy change' to facilitate collaboration with other domestic agencies on water issues, as well as to cope with the globalizing aspect of the issues. Another respondent commented that, although there appear to be some changes (within the Ministry of Transport and Ministry of Agriculture, Forestry and Fisheries) after WWF3, he is not certain whether such changes were generated because of WWF3 only, or there had been an ongoing trend earlier and the conference has served as a trigger to materialize such changes.

Other post-WWF3 changes mentioned by some of the respondents were the same as the answer to Question 11: New initiatives.

8.3.10 Question 10: Changes in Investments Due to Megaconferences

In the original questionnaire, this question was posed as follows:

> Has the investments availability for the water sector in your country increased or decreased by these conferences, which would not have occurred unless these conferences had taken place? If so, please provide the direction and a rough estimate of these changes. Or have these events had no perceptible impacts on investment availability in your country?

This question was slightly modified in the Japanese version as follows (answers follow):

> Question 10a: Do you think the investments availability for the water sector (including public works, business activities, overseas development aid, research activities) in your country have increased or decreased by these conferences, which would not have occurred unless these conferences had taken place? (Twelve answers to the original questionnaire are also incorporated here.)

Result:
Yes: 13 respondents (22.8%)
No: 14 respondents (24.6%)
No idea/No answer: 30 respondents (52.6%)

> Question 10b: If you have answered 'Yes' above, please specify the perceived changes in the investment availability.

Most of the answers to this question were not specific without any concrete project names or rough estimate of changes. Eight respondents said that there have been some changes in the amount of budget allocated for water-related activities, as well as in the number of water-related projects including research activities. One respondent mentioned the establishment of the new research project "Modelling and Utilization System of Water Circulation" under the Core Research for Evolutional Science and Technology (CREST) Programme of the Japan Science and Technology Agency (JST) as an example of a new investment, the impetus of which arose with the increased understanding of the importance of water issues, though he also noted that it was not because of the influence of the megaconferences only. Some of the other answers were in common with the answers to Question 11: New initiatives (i.e. "Clean Water for People" initiative by Japan and the United States following UNCSD, the establishment of the UNESCO Water Research Institute) (see below).

Two respondents mentioned changes in the amount of water-related overseas aid, which was at least partially influenced by WWF3. One of them stated:

> With WWF3 and previous commitments made to the water sector in developing countries, water issues are now one of the few prioritized areas in the recently modified Japanese ODA (Overseas Development Aid) Charter. In this sense, despite financial deficit, Japan will try to maintain its high-level financial support to the water sector in developing countries, though the figure is not certain.

One respondent commented, on the contrary, that although there seem to be some qualitative shift in water-related development projects, quantitative changes in the amount of investment have not been seen.

8.3.11 Question 11: New Initiatives, Including Water Sector Reforms

The question here was asked as follows:

> In your view, did some new initiatives originate from these events, which otherwise would not have occurred? What are these initiatives? Also, did these events contribute to water sector reforms in your country? If so, in which areas?

Answers are listed below as presented by the respondents (by a single respondent each unless otherwise specified):

New NPO activities and networks

- Establishment of the Japan Water Forum (NPO) following WWF3 (six respondents)
- International Network for Water and Ecosystems in Paddy Fields (INWEPF) (three respondents): established by the Ministry of Agriculture, Forestry and Fisheries and its partners in November 2004 following the results of WWF3, as an international network to discuss and disseminate information on multiple functions and roles of paddy farming.
- Network of Asian River Basin Organizations (NARBO): established in February 2004 by the Asian Development Bank (ADB), ADB Institute and Japan Water Agency for the purpose of capacity building of River Basin Organizations (RBOs) in Asia to promote Integrated Water Resources Management (IWRM).
- Youth World Water Forum: a network of the world's youths was established following WWF3, adopting a declaration and action plan. The network will hold "Youth World Water Forum" at the Aichi Expo in August 2005.
- Rainwater utilization network for NPOs, local governments and businesses were born after WWF3 (activities: rainwater utilization seminar for architects held in Tokyo in spring 2004; on the basis of the rainwater harvesting session of WWF3, Tokyo Asia Pacific Sky Water Forum to be held on the theme building international networks towards solving the water crisis in Asia, August 1–7, 2005, in Tokyo).

Research initiatives

- Global Water Cycle Research Initiative under the Council for Science and Technology Policy of the Cabinet Office.
- Establishment of the UNESCO Water Research Institute (tentative name) under the Public Works Research Institute.
- Some new water-related research initiatives aimed at the creation of a recycle-oriented society within the CREST Programme of JST.
- Research initiatives on Blue Revolution and Water Governance.

Other government initiatives

- Japan–United States "Clean Water for People" initiative (two respondents): launched in June 2001 based on the discussion at UNCSD.
- "Japan–France Water Sector Cooperation": launched in March 2003, following UNCSD and WWF3.
- Water Environment Partnership in Asia (WEPA): initiated by the Environment Ministry.
- Establishment of the International Department within the Japan Water Agency.
- Establishment of the Sewage Globalization Committee within the Sewage Department of MLIT.
- Increased collaboration between government agencies as the understanding on IWRM has prevailed (three respondents).

Three respondents commented that there have been no new initiatives within the country, while one commented that no clear observation can be made at the moment as a genuine reform takes a long time.

8.3.12 Question 12: Key Lessons

The question asked here was as follows:

> What in your view are the key lessons (positive and negative) that we can learn from these megaconferences?

Many of the answers to this question, both positive and negative, were the same as the ones for Questions 4 and 5 on strengths and weaknesses of megaconferences, namely:

- Megaconferences contribute to awareness raising, fostering common understanding and information exchange (nine respondents).
- There is no use or significance as there are no concrete policy implementations or actions following as outcomes (five respondents).
- Need to clarify discussion themes and make the conference smaller scale (three respondents).
- It is important to minimize the cost of organizing such conferences (two respondents).
- Low impact on the general public despite the participation of state leaders (one respondent).

Other comments include (also partly in common with answers to the earlier questions):

- Domestic water community has been 'internationalized'.
- Having seen the decision-making process at the megaconference, I realized that the conventional decision-making process of Japan is outdated today.
- Megaconferences can serve as a place to extract political commitment of the participating governments, but not enough evaluation has been done as to

whether such commitments have yielded any practical results or not. At least megaconferences play the role of providing a discussion framework which is particularly important for a field where there is no entity of international or global governance. It is important to move discussion forward step by step through a series of such forums.
- Bandwagon effect?—Some countries or organizations might feel the need to participate in such megaconferences simply not to fail to jump on the bandwagon.
- There is a certain limit in what megaconferences can do. For instance, such global conferences do not seem to be a suitable place for discussing issues which are of more regional concerns rather than global. On the other hand, by bringing such issues to megaconferences, and having them ignored or unfocused on such occasion, there is a danger of losing the significance of the issues just because they were not taken up seriously at the megaconference.
- Discussions and studies are needed as to how to organize such conferences more efficiently without too huge a budget and logistical burden.
- It was learned that such megaconferences are navigated by those who are "good at" speaking up at such meetings, while there is another group of people who do the actual work but are not good at negotiations or diplomatic socializing.
- I have learned that the conflict between developing and developed countries is the most important problem, as in other international issues.

8.3.13 Question 13: Changes in the World of Water Due to Megaconferences

The question asked here was as follows:

> In your view, would the world of water been any different if these conferences had not taken place? If you think the world has changed, in what ways has it changed?

In the Japanese version, this question was asked as (answers follow):

> Question 13a: Do you think the current situation of water resources or water sectors of the world would have been any different if these conferences had not taken place? (*12 answers to the original questionnaire are also incorporated here.*)

Result:
Yes: 27 respondents (47.4%)
No: 10 respondents (17.5%)
No idea/no answer: 20 (35.1%)

> Question 13b: If you have answered 'Yes' above, in which ways has it changed?

Answers given to this question were as follows. Again, many of them are in common with earlier answers.

- Awareness and understanding about water issues have increased, water discussions globalized (15 respondents). One of them commented: "Ways to look at water issues have changed from a one-size-fit-all plan of water development to tailor-made water management considering economic, social and cultural aspects in an individual country. It is definitely a move forward."

- Changes in the field of development aid: increase in water-related projects, generating consensus on the water-related development strategy among donor agencies; improvement in the water situation in certain focused geographical regions (although, as one respondent noted, megaconferences do not function as a mechanism for resolving water conflicts) (five respondents).
- Megaconferences have contributed to the formation of certain water-related international networks which lead to more issue-specific discussions and concrete actions.
- Widely introduced new water-related concepts such as 'virtual water' into the world.
- "Big water supply companies have succeeded in strengthening their power in the world water resources stage."

8.3.14 Question 14: Overall View on the Megaconferences

The original questionnaire asked:

> What is your overall view of these megaconferences? Which of the following statements come closest to your view?
>
> A. The global megaconferences are useful and cost-effective. We should continue with them, but with only marginal changes.
> B. The conferences have now become one big "water fair", with a lot of activities but without much thought as to their relevance, appropriateness, outputs or impacts. There is no coordination between events, no clear focus, and their cost-effectiveness leaves much to be desired.
> C. The concept of such global conference is good, but the present framework for organization needs to be changed radically. The events should be more focused and output-oriented. The main criteria for success should not be the number of people who attended the conference, but the quality of the results and their impacts.
> D. Instead of the global megaconferences it would be desirable to organize regional meetings, dealing with regional problems and issues, and which could be focused and impact-oriented.

In the Japanese version, the comments "E: Other than above. (Please specify.)" and "F: I don't know" were added. A few respondents selected more than one answer. The results are shown in Fig. 8.8.

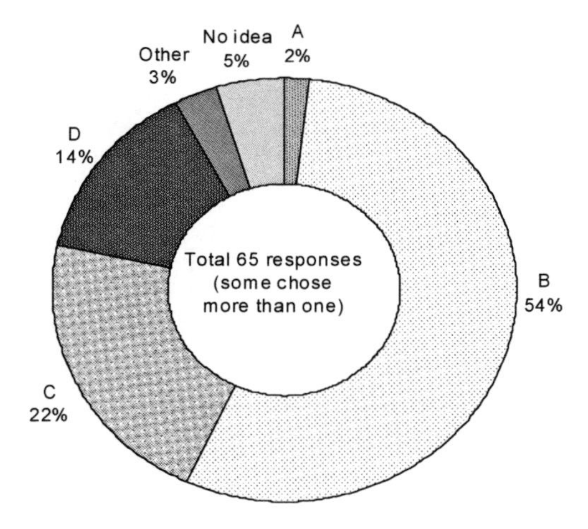

Fig. 8.8 Answers to Question 14

Comments of those who marked "E: Other than above" are as follows:

- Although it might be a rather optimistic view, I feel that megaconferences play a role of getting political commitments from the governments. Each megaconference should not be evaluated from the viewpoint of cost-effectiveness, but we should also see the series of the conferences as a long-term process of generating faith and trust among participating states, which enable them to act in harmony on certain groups of themes through the repeated ceremonial meetings. (This respondent also marked B and D.)
- The existing discussion forums of various organizations should be more utilized instead of holding megaconferences. When megaconferences are needed, the practicality and meaning of such conferences should be seriously considered beforehand. Follow-up activities of megaconferences should be conducted by existing organizations rather than establishing new ones. There should be more than two decades of interval between megaconferences of the same theme.

8.3.15 Question 15: Ministerial Declarations

The question was posed here as follows:

> At many of these conferences, there were Ministerial Declarations which had relevance to the water sector. Please give your opinions on the following questions:

In the Japanese version, the following sub-questions were slightly modified from the original questionnaire (answers follow).

Question 15a: Are you and your colleagues aware of these Ministerial Declarations? If so, which ones? *(12 answers to the original questionnaire are also incorporated here).*

Result:
Aware of the existence of any of them: 42 respondents (73.7%)
Of them, aware also of their contents: 27 respondents (47.4%)

Aware of the Ministerial Declaration(s) of:
WWF3: 34 (81.0%) (% out of 42)
UNCED: 16 (38.1%)
UNCSD: 14 (33.3%)
WWF2: 8 (19.0%)
WWF1: 3 (7.1%)
Bonn: 3 (7.1%)
Most of the Ministerial Declarations: 3 (7.1%)

Question 15b: If you are aware of the Ministerial Declarations, please identify the conference whose declaration you consider the best and had the most impact.

Question 15c: What are your views on the relevance, appropriateness and usefulness of such Ministerial Declarations?

Answers for the above two questions are as follows:

Mal del Plata (One Respondent)

- The conference [not necessarily the declaration] itself stimulated in a deep sense the public concern and interest on water for the benefit of the future of mankind; at least, it gave a certain impact to everybody who served in the water issues, particularly among the people who were engaged in the planning sector.

UNCED (Thirteen Respondents in Total)

- Showed the political commitment of the governments, and generated impacts on domestic policies and overseas development aid (whether the actual implementation followed or not) (seven respondents).
- Contributed to wide awareness raising on such global issues as sustainable development, global warming, biodiversity, etc. (three respondents).
- Though very slowly, it contributed to expansion of international laws (irrespective of their effectiveness).

WWF2 (One Respondent)

- It brought the gender issues to the forefront.

UNCSD (Four respondents)

- It balanced economic, social and environmental aspects well.
- It set the direction that the international community should head for.

WWF3 (Five respondents)

- It set the direction that the international community should head for.
- Publicity effect to raise awareness on the importance of issues.
- Showed the political commitment of the governments, though it is not legally binding.

Other comments on the significance and role of Ministerial Declarations:

- It shows political commitments (three respondents).
- Publicity: it clarifies the messages from the conference (two respondents).
- It clarifies the imminent tasks; also the process of drafting the declaration may be useful in understanding each other's position (two respondents).
- It is better than not having any declaration (one respondent).
- There is no significance, role or impact (eight respondents): "They are little more than ceremonial addresses."

Question 15d: If you are aware of the Ministerial Declarations, do you think the domestic water policy and/or priorities of water programmes have been affected by them?

Result:
Yes: 12 (21.0%)
No: 14 (24.6%)
No idea/No answer: 31 (54.4%)

Question 15e: If you have answered 'Yes' above, please briefly provide example(s).

The answers shown to this question seem to rather overlap with answers to the earlier question on the impact of megaconferences on domestic policies:

Impact on domestic policy, laws and water-related projects (six respondents):

- By highlighting the outcomes of the conference in the form of a Ministerial Declaration, it serves as an indirect incentive to start projects to create new research initiatives and policy suggestions.
- Lead to the promotion of environmental policies, environmental legislation, adoption of new environmental projects.
- UNCED Declaration → Convention on Biodiversity → National Strategy on the Conservation and Sustainable Use of Biological Diversity (1995) → Law for the Promotion of Nature Restoration (2002), Law for the Prevention of Arian Species (2004).
- Introduced the concept of 'sustainable development' as a basic principle at the bottom of the country's water-related policies.
- Influenced the initiative of the Tokyo International Conference on African Development (TICAD), and establishment of other policies in accordance with the Millennium Development Goals (MDG).
- Increased the general understanding of water issues among decision-makers.

Impact on development aid policies (two respondents):

- Influence on the ODA policies.
- Increased budget allocation for the environment (due to UNCED, UNCSD) and water (due to WWF3) sectors.

8.3.16 Question 16: Expectation for Future Conferences

This question was added in the Japanese version of the questionnaire. The question was posed as follows:

Question 16a: If conferences on water are to be held on a global scale in the future, what do you expect about/from them?

Many of the answers to this question reflected the respondents' views on the strengths and weaknesses of megaconferences as expressed in the answers to Questions 4 and 5 above.

- Should clarify the theme and lead to concrete actions and outcomes (14 respondents).
- Should contribute to awareness raising, common understanding and clarifying imminent problems and tasks (five respondents).
- Should serve as a place for information exchange on various water issues and latest technologies (five respondents).
- Should be regionalized, grouped according to themes, and held on a smaller scale (two respondents).
- Should provide various stakeholders (especially those in the weaker position in the society) with opportunities to speak up (two respondents).
- Should continue to serve as a discussion forum (two respondents): "It is important to have concrete action proposals, but it is also important just to have an opportunity to hear different opinions from people in various positions and backgrounds. Detailed discussions are not possible at such megaconferences anyway, so it is now better to consider them simply a place to hear others' opinions".
- Should minimize the cost as much as possible (two respondents).

Other expectations, expressed by a single respondent each, include the followings:

- nurture personal/professional networks;
- clarify the future direction of water resources management;
- increase the government's support for developing countries in short of basic infrastructures;
- lead to the collaboration between the public sector, academics and the private sector;
- edify the mass media;
- serve as a place to feel the world trend on water; and
- do not end up as a one-off event but continue the forum for building trust.

Also, one of the respondents expressed his view as follows:

> After all, WWFs led by the World Water Council, which is dominated by water-related businesses, are not democratic, and the conferences are used simply to appeal for more large-scale investments and there are no discussion to support those really suffering. Ministerial Declarations are not legally binding, so they have little use. Personally, I think that the discussion on water should be continued within the United Nations Commission on Sustainable Development. The United Nations should take the initiative on water issues, and for that purpose the United Nations Water Charter supported by 24 United Nations agencies concerned with water should be adopted. Also, the issue of whether water is a public good or not should be clarified within the international community.

Question 16b: Would you like to participate in water-related megaconferences in the future? Please state the reason.

Result:
Want to participate: 27 respondents (60%)
Do not want to participate: 5 respondents (11.1%)
I don't know/No answer: 13 respondents (28.9%)

Reasons for wanting to participate (some gave multiple answers) (by one respondent each unless otherwise specified):

- For information gathering and interaction with other fields (11 respondents).
- May be useful for my work (business or NPO) or research activities (four respondents).
- To feel and understand the world trend on water issues (four respondents).
- To increase supporters for my organization's activities (like on the occasion of WWF3).
- For career building.
- For job hunting in the water-related field.
- To learn about the procedure, customs and trouble-shooting at international meetings.
- For personal/professional networking.
- To disseminate information from our side.
- Because I have never participated in the past.
- To see whether the conference proceeds in a more democratic manner compared with the earlier WWFs.

Reasons for not wanting to participate:

- Waste of time and money if conferences continue to be a sheer merrymaking for the governments and businesses (four respondents).
- Language barrier (simultaneous interpretation cannot be trusted) (one respondent).

Out of the five respondents who clearly said they would not like to participate in megaconferences in the future, three had never taken part in any of the megaconferences targeted in this survey, while the other two had participated in WWF3 but not other conferences.

8.3.17 Question 17: Comment

Please give your views on any other aspect(s) and issue(s) of global megaconferences not mentioned above.

Twenty-seven respondents gave comments of some sort. About half (13) were negative comments expressing the respondents' sarcasm or suspicion regarding the meaning, substance, process, cost-effectiveness and fairness of megaconferences, which were already expressed as answers to other earlier questions though in different words.

There were also positive comments though much smaller in number, again in line with what was mentioned as answers to earlier questions already. One commented that, unlike the conventional bilateral or multilateral governmental meetings, the fact that citizens' groups (NPOs) can be regarded as a main actor at megaconferences should be positively valued. Also there were views that the continuous effort to have such forums may be recognized to have some positive value, since some intergovernmental, intercultural or interpersonal faith and trust may be bred through a series of such global forums, even if realistic solutions to all the problems could not be born right after each conference.

One respondent mentioned the need (for the Japanese government) to constantly secure capable manpower to cope with the huge logistical burden and to steer a wide range of discussions at such megaconferences, whether the conference is to be held locally or not.

A few respondents commented on the method of evaluating megaconferences itself. The most notable comment that the authors felt should be cited for a later discussion is the following:

Many people will be interested in this sort of survey trying to see the effectiveness of megaconferences, but it is highly dangerous to make judgement on the significance of each conference based on the views expressed through this questionnaire only. [...] I would like to expect that this survey will contribute to the development of a new methodology to evaluate the effect of megaconferences, rather than to come up with a certain conclusion regarding the outcomes and impacts of the past megaconferences right away.

8.4 Concluding Remarks

As stated at the beginning, it is not the intention of this report to draw any conclusion from the findings of this questionnaire survey in Japan. In addition to the small number of responses received, the extent of the respondents' experience and knowledge was (somehow inevitably) biased towards WWF3. This conference was held and most widely publicized in Japan, and is at the moment of writing the most recent (therefore probably most remembered) megaconference on water furnished with a Ministerial Declaration. It was also exceptionally large[1] for a

[1] According to the WWF3 Secretariat, over 24,000 people took part in WWF3 while the Secretariat had earlier expected the number of participants to be around 8000. The number

conference exclusively on water issues. The respondents' views expressed in this survey may largely reflect what they perceived about WWF3, and may or may not necessarily be used as a clue to judge people's views on other conferences. (The authors of this report also have the experience of participation in WWF3 only, among the megaconferences targeted in this survey.) Also, there may be a need to see the entire series of megaconferences as one trend,[2] rather than trying to judge each one of them separately. It is not known how many of the respondents assumed such a viewpoint if at all. Also, there were differences in what people perceived as 'effect' or 'impact' of the conferences, and many of the respondents seems to have mixed opinions about their positive and negative aspects. A majority of the respondents seems to feel that the conferences were too expensive and extravagant, and discussions were unfocused and concrete outcomes and actual impacts are difficult to see. However, many of them did not dismiss the idea of holding such conferences, positively wanting to participate in future conferences of the same or a similar nature.

In terms of the methodology of this global survey, the authors felt that there might have been some, potentially critical, limits. This was pointed out by some of the questionnaire respondents as well as at the workshop held in Bangkok in January 2005,[3] where survey results from various parts of the world including Japan were presented.

Firstly, the eight conferences on water targeted in the current survey could be classified into different categories according to the nature of each conference, which in turn might have implied different expectations and outcomes. Specifically, they can be divided into those focused specifically on water and those more broadly about environment and development issues, or those initiated by the United Nations and those organized by non-United Nations bodies, and/or those which took up a certain topic for the first time and those which were follow-up meetings of other earlier events. Then, there is the question of whether it is justifiable enough to group these conferences together and render to simple comparison.

Secondly, as discussed above, it might be necessary to view the series of the conferences as a trend leading to some movements, conceptual shift or institutional reform as a result, rather than trying to evaluate each and every conference separately. In the questionnaire used for the current survey, the approach to lead the respondents to assume such a view might have been rather low-key. Also, during the workshop it was pointed out that it might be necessary to observe each conference

of overseas participants only amounted to some 5800, which is already more than the total number of participants in WWF2 (Matoba 2003).

[2] For instance, Ishimori (2004) tried to examine how the series of international agreements on water have affected the scale and content of water-related overseas development aid projects by developed countries and international organizations, using the Creditor Reporting System (CRS) of the Organization for Economic Cooperation and Development (OECD) Development Assistance Committee.

[3] Workshop on Impacts of Megaconferences on Global Water Development and Management, Bangkok, Thailand, 29–30 January 2005, sponsored by Third World Centre for Water Management, Mexico, with support from Sasakawa Peace Foundation, USA and Japan.

as a longer process including its preparation period in order to grasp more specific impacts and changes following the conference.

Thirdly, as an afterthought of attending the aforementioned workshop which provided us with the first and most interesting opportunity to discuss this global survey with various experts, the authors felt that it would be probably unrealistic to assume that impacts of the 'global' water conferences should spread and be observed uniformly at every corner of the world, like the ripples caused by a raindrop onto a still water surface. Whether certain impacts of a global conference can be felt or materialized in a country or not might depend on the political, socio-economic (e.g., development status), and/or cultural (e.g., languages, different culture of decision-making, etc.) conditions of the country. The pace and the ways in which the impacts appear might also vary due to such domestic conditions. For instance, in the case of Japan, information on earlier (especially pre-internet-era) conferences which were mostly conducted in English or some other European languages might not have been shared within the country so much as in other developed countries. Also, the public interest in environmental issues started growing within Japan much belatedly compared to other countries in the West, and such delay also might have contributed to the different pace of this country to initiate some changes in accordance with the global megaconferences on environment and/or water. At the Bangkok workshop it was reported that many of the respondents of the global questionnaire survey evaluated the United Nations Water Conference in Mar del Plata (1977) more highly than others, but that was far from the case with the Japanese survey, which showed rather low awareness of even the fact that such conference had been held. Therefore, as some of the respondents of the Japanese survey commented, it might be even still too early to observe concrete impacts within this country. If that were the case with Japan, the same might apply to some other, especially non-Western, countries.

Such considerations about the basic approaches and assumption of the global survey might be useful for elaborating the methodology and for more in-depth discussions of the survey results in the future. In any case, it was clear that many of the respondents of the Japanese survey, as well as those of other country surveys, agree in that "both the procedure and substance of the conferences need to be given a serious review" (comment from one Japanese respondent). This "serious review" is exactly what the global survey initiated by the Third World Centre for Water Management, which this Japanese survey tries to add to, is intended for. The number of responses in the Japanese survey was rather too small to enable any meaningful quantitative or qualitative analysis. However, it is hoped that, together with the survey findings from other regions, it will contribute to some meaningful discussions regarding the future direction of water-related megaconferences. Also, as one of the respondents commented, while it is dangerous to judge the significance of megaconferences with this type of survey only, it would be a great contribution of this global study if even a small step is made towards the development of a new methodology to effectively evaluate the impacts of water-related megaconferences.

References

Ishimori K (2004) Water Issues and International Development Aid. In: Development Approaches and Changing Sector Priorities. Foundation for Advanced Studies on International Development (FACID) pp. 215–242 (Japanese)

Matoba Y (2003) Some Outstanding Features of the Third World Water Forum. Journal of the Society of Irrigation, Drainage and Reclamation Engineering 71(7): 22–25 (Japanese)

9 Impacts of Megaconferences in India

C.D. Thatte

9.1 Introduction

On request from Mr. M. Gopalakrishnan, Secretary-General, about 30 responses were received by ICID Central Office located at New Delhi, to the Survey Questionnaire circulated earlier by the Third World Centre for Water Management. Out of these responses, about 23 respondents had participated in some of the listed conferences; others were associated either with the process leading to the conferences or with application of the outputs of these conferences during their work. Mar del Plata Conference was attended by about 40% respondents, 1st World Water Forum (WWF) by 50%, 3rd WWF by 80% and the rest by about 65% of respondents. The sample of responses is therefore considered fairly reliable. A summary of the responses received to the questions posed in the questionnaire is compiled and presented in this chapter.

9.2 Overall View About the Events

The responses are graded from Poor assigned '0' rank, increasing to Excellent given rank '5'.

Rank 3, average, was given to most of the events. Ranks 4 and 2 followed. Poor and excellent ranking was given by only 10% respondents. The overview responses thus indicate a normal distribution around 'average' which means a 'so-so' view.

9.3 Impacts (Ranking A for Excellent, Going Down... Up to F for Forgettable Impacts)

A highly varied response was received. The majority ranked impacts on C and D groups, again indicating an average impact. Nearly 20% respondents graded impacts as F which should be of concern.

9.4 Strengths

The events received international and media attention and they created significant awareness amongst participants and the general public attentive to media on such issues. For those interested, access to new ideas on the water sector was certainly facilitated.

The events facilitated interaction of like-minded people. The environment sector's 'voice' was undoubtedly strengthened. The events have clearly started facilitating co-existence of 'development and environment' concerns in the same conference and at times on the same platform. Those voices which were *hitherto* starved of funding could explore new options for funding. The possibility of influencing government policies was enhanced.

Some events brought out in advance 'issue papers'. They were widely circulated and posted on dialogue websites. Responses were collected and synthesis attempted. It was to be used for consideration during the events, which possibly did not happen defeating the pious motive. It is a very good concept which needs to be adopted in the future.

Some respondents, however, did not find any 'strengths' in these events and considered them a big waste.

9.5 Weaknesses

Time available for discussions was considered too short by some. Events themselves were considered very expensive.

There were too many sessions and themes causing loss of focus. There was considerable over-emphasis on 'environmental' concerns as against 'human development' issues of survival, hunger, thirst and energy. For instance, agriculture which accounts for 70% of global water use was sidelined. Many sessions were hijacked by lung-power of activists who were provided liberal financial support by donor agencies and other vested interests for travel and expenses towards attending the events. Possibly in order to show concern for activists, real workers/ professionals got ignored in providing financial support. Activists aimed at fomenting controversies, rather than building bridges.

The few 'issue papers' circulated in advance, were largely drafted by individuals who typically highlighted views of the developed world or were donor-driven rather than demand driven. The presentations lacked requisite exposition of both sides.

The much desired integrated views did not emerge from the events. One got only a kaleidoscopic picture. It was seen that diametrically opposite views/conclusions emerged from adjoining session halls. Was disharmony desired? Of course both sides got the dubious satisfaction of making their points. The question remained – For whose benefit? Representatives from developed countries, coming from funding agencies and as such working as organizers, dominated the outputs which were tailored to please certain activists from developing countries opposing development.

Some outputs were good but were not in an 'implementable' format. Enough lip sympathy to the causes was provided by government representatives but they were neither hopeful nor enthusiastic about their adoption back home, citing reservations and constraints.

Responsibility for dissemination of outputs up to the right levels was not specifically assigned and hence it remains weak. Multiple venues were distractive and proved counter-productive.

9.6 Cost-Effectiveness

The unanimous view of the respondents was that the events were not cost-effective. They need to be improved drastically. For instance, the kick-off meetings were also needlessly held on a grand scale without being useful except of course serving publicity for the funding agencies and their sponsors. Events served more as fun-fairs; they had a dispersed focus, size of participation was unwieldy, causing dissipation and dispersal of effort.

Cost would have been immaterial, if benefits flowing from the outputs for the needy and deprived societies had been commensurate. Seemingly the outputs only benefited self-appointed experts coming from rich societies.

Some felt that the size of funding for these events could have been better used on providing the much needed infrastructure in some needy societies.

9.7 Documentation: Adequacy, Quality, Dissemination

Quantity and quality generally were considered adequate. If one looked for distilled wisdom, it was not available at the end of the day. Dissemination of outputs in needy countries was found inadequate. Recommendations from successive meets needed a review, compilation, comment, mid-course corrections and dissemination in local languages. WWC-GWP-IWALC could handle such follow-up better, if part of the funding for the events is reserved for it.

Some found the output not reflecting (and disregarding) real needs of developing countries. Resolutions were not available even with policy-makers, because their relevance was doubtful. One respondent suggested assigning dissemination to participants.

9.8 Were the Results Positive, Practical and Implementable?

Little discernible results were available, apart from increase in general awareness about global status. The results in a way highlighted 'lack of congruence' on critical issues such as availability and variability of waters, need for more dams, international shared river waters, environment issues, etc. Critical issues like food insecurity were sidestepped. Results were found to be marginally applicable at regional and/or national level. If the events were aimed at developing consensus, the effort undoubtedly failed. On the contrary, raging controversies were fuelled.

The World Water Council as the main promoter should be the custodian of the results of these mega events, providing a clear perception of issues needing consensus development, their practicability, etc. for future follow-up.

Some of the professional organizations (like ICID, for instance) pondered over the output and developed their own strategy for implementation of outputs of 'Vision'. Such efforts were not even recognized nor supported in later events, as if each event is a 'stand-alone' one, which unfortunately is not true. A continued thread of these events needs to be apparent in the organization.

9.9 Changes in National Policy

In general there was a varied response. A minority found that there was a slow but sure influence in policy evolution. The majority felt otherwise, as they found that the differences being experienced at home were fiercer than at global level. Also that the events did not aim at finding rational course, nor at influencing/correcting wrong views.

Also, they felt that even ministers/officials who participated did not use these outputs for changing policies at home. No continuing mechanism was available nor aimed at as an output for making such changes.

National priorities prevail, not outputs of mega-meets. There was no dialogue to assess synoptically the critical needs of at least the 'hot spots'. The tendency amongst national leaders is to use only such outputs that suit them, disregarding the rest.

9.10 Did the Events Result in an Increase or Decrease in Investment in the Water Sector?

Most felt that the events exercised little or no influence. In reality, a continuous decrease in 'own investment' in the water sector as a result of competing demands in other sectors is apparent. The influence of multilateral and bilateral funding agencies in the national water portfolio was considered small and found wanting in taking 'hard' decisions like supporting infrastructure development due to their vulnerability to activism. For instance, the changes on 'dam policy' of the World Bank came under scrutiny from both sides of the controversy, as never before.

Globalization/FDI processes were of recent origin. It was too early to see their influence on these processes. Financial support to NGOs (mostly anti-establishment) in particular from West Europe was significant and was seen to have multiplied creating difficulties for pursuit of even accepted policies. For some, it was evidence of interest of big business in disallowing developing countries from growing stronger economically.

9.11 New Initiatives/Reforms Set in Motion Owing to These Events

New initiatives were not really set in motion. The events, however, highlighted 'non-structural (soft)' issues in preference to 'structural (hard)' issues, owing to which many countries had made significant progress in the water sector.

Some of the non-structural issues were not necessarily progressive and caused obstacles in (delayed) ongoing activities. Buzz words like 'sustainable development, transparency, participatory approach, human rights, water markets, virtual water... etc.', became the key elements in stopping whatever little was being achieved. Issues such as over-exploitation of groundwater (in absence of availability of surface waters), environmental concerns, micro-watershed development, local actions, low water use efficiency for irrigation and hence productivity, full cost recovery and pricing, institutions and governance, corruption in the water sector etc. have gained momentum rightly or wrongly, sometimes ignoring the basic issues, leading to sustained under-development.

9.12 Key Lessons

Lessons have been culled from the responses received and are listed below in two specific groups.

1. Water sector: Mass awareness, stakeholder participation. First think globally—then locally—and then one can act quickly and locally to serve home needs, situations and reality. Although ministers/officials participate, their interest is short lived for the period they occupy certain positions. Often, they use outputs selectively to support their own political agenda. There is a need to integrate and build bridges between developmental and environmental concerns. Controversies should not be fuelled, and their futility should be highlighted. Global solutions are not practicable. Success stories should be publicized after scientific assessments, and failures should not be concealed.
2. Mega-events: The fun-fair ambience should be eliminated. There should be separate awareness building and policy orientation aspects. The latter has to be

aimed at addressing similar groups of countries – according to socio-economic status, water availability, development needs, management status and unmet needs status. The focus should be narrowed, avoiding too many themes/sessions. The processes of Forum (providing opportunity to voice) and Policy (rational, needed one, practicable, realistic, supported by Science and Technology) need to be separated. National, then regional, then global consensus should be built where possible. Global commitments are dilute or are pedantic. National action plans and not pious policies are essential. The events have become repetitive.

9.13 Has the World Changed as a Result of These Events?

There was a varied response, ranging from no discernible change to significant change. However, the consensus was that the events effectively brought out the complexities of the water sector. Some felt that view of the 'developed world' had changed (for better or worse – no clear opinion) over the three decades. Having accomplished water-related tasks, they lobby 'change' (building pressure) through financial clout, offering overseas training, offering (and employing) services of often inexperienced own consultants in the 'developing world' through NGOs, adding to the conflicts and controversies and delaying solutions of 'water'-related problems. Such 'large' negative influences were also apparent at mega-events. Unfortunately, it eclipses the 'small but positive' change possible as a result to these events.

9.14 Utility of the Events

The approximate proportion of responses was as follows: A – Useful (10%), B – Leave much to be desired (20%), C – Change focus radically (40%), D – Go for regional meets instead of global ones (30%). The latter two figures indicate that the consensus was in favour of holding regional meetings with changed focus to suit their needs.

9.15 Ministerial Declarations

- Awareness: about 40% of respondents were aware of these declarations.
- Relevance, etc.: about 50% felt that the events were relevant, others felt they were not (50:50).
- Effect on country policy: about half of the respondents felt that there was little impact of declarations on the country's policies, the other half felt that there was almost no influence.

9.16 Views on Other Aspects

Responses mostly covered the familiar views. Some specifics were as follows: country water vision, policy, action plan, etc. have to fit in certain minimum parameters of global vision. However, the Least Common Denominator (LCD) was elusive. Policy-makers often do not insist on implementation, instead they use policies for serving political agenda. The developed world should not support participation of NGO representatives to serve their own agenda, ignoring the need for participation by professionals who are at the cutting edge. The funding agencies should not build activism as a career in the developing world. Local actions must fit into the basin/national policies/plans, and should not be harmful to each other, or appropriate resources. Water is a public national asset, not a good. They should avoid over-emphasis on the environment, ignoring the development needs of the poor. Supply management is as important as demand management for developing countries, owing to the rapidly growing unmet needs. Promotion of development should not be despised nor discouraged. Targets to satisfy MDGs should be the first aim. Each country must develop consensus on action plans to serve MDGs. Then they should integrate them regionally, in particular for shared river basins. Rather than concentrating on 'local actions' for the next WWF, feedback should be sought from ministers of the MDG countries and woven into the agenda. Megaconferences should not become mega-touristic events.

10 Megaconferences: View from Bangladesh

ATM Shamsul Huda

10.1 Introduction

In the last quarter of a century, a number of megaconferences on water and environment have been held to address the burning issues of development of the day and recommend solutions to some of the identified problems. There is, however, persistent criticism about the efficacy and cost-effectiveness of such conferences, though the views expressed on them are mostly conjectural in nature. The Third World Centre for Water Management in Mexico has undertaken the rather unpleasant task of evaluating the impact of such megaconferences on water management at regional/national level primarily on the basis of perception of the regional/national water management community with the implicit aim of improving the standing and performance of similar events. The evaluation would be based on a global-scale survey of the opinion of concerned people and for this purpose a structured questionnaire was prepared and circulated by the Centre. Among others, Bangladesh was also selected as one of the target countries for the purposes of this survey. This report seeks to collate and analyse the responses received from the respondents as part of that global exercise.

Bangladesh Unnayan Parishad (BUP), a national think tank, was given the responsibility of circulating the questionnaire and analysing the responses. The questionnaire circulated to the Bangladeshi participants was the same as the one sent by the Third World Centre and no changes were made. The following eight megaconferences were included for the survey:

1. United Nations Water Conference, Mar del Plata, 1977.
2. International Conference on Water and the Environment, Dublin, 1992.
3. United Nations Conference on Environment and Development, Rio de Janeiro, 1992.
4. First World Water Forum, Marrakech, 1997.
5. Second World Water Forum, the Hague, 2000.
6. International Conference on Freshwater, Bonn, 2001.
7. United Nations Conference on Sustainable Development, Johannesburg, 2002.
8. Third World Water Forum, Kyoto, 2003.

For purposes of analysis, the conferences are grouped into three categories. The United Nations conferences (Mar del Plata, Rio and Johannesburg) are grouped in the first category for the reason that they are the highest level forum for enunciation of policy at the global level under the United Nations System and have some influence on their follow-up at the level of the national governments.

The international conferences (Dublin and Bonn), largely participated by technical people and various stakeholders, seek to thrash out the policy issues that need to be placed before the United Nations conferences. The World Water Forums (Marrakech, the Hague and Kyoto) are basically geared to dialogue among different stakeholders, information dissemination and raising awareness on different sensitive water issues. Ministerial Declaration is an essential part of the Forum proceedings but it does not carry the same weight as in the case of such declarations under the United Nations System. The reason for making the above categorization is that with the exception of two respondents out of 15, none had attended either the United Nations or the international conferences. The overwhelming majority of the respondents have answered the questionnaire based on their experience of attending either one or two of the World Water Forums. While reading through their responses, it would be useful to keep this limitation of the survey in mind.

10.2 Survey Findings of the Megaconferences

10.2.1 Participation

The BUP prepared a list of 21 individuals for soliciting their opinion that represents a broad array of interests and specializations – academicians, water experts, government officials, NGO representatives and civil society advocates. Given the history of poor response to these types of surveys in this country, the BUP decided to personally contact the targeted respondents and later on collect the completed questionnaires. Of the targeted 21, 15 respondents returned their completed questionnaire, which represent a good percentage for the purpose of analysis. The majority of the respondents did not mind to disclose their identities while three respondents preferred to remain anonymous. The break-up of participation by category of conferences attended shows the following:

Not participated at all	2
Participation at United Nations Conferences	2
Participation at International Conferences	1
Participation in World Water Forums	19

None of the respondents had attended either of the United Nations conferences of 1977 and 1992. Though none had attended the First World Water Forum, eight had attended the second one and of these eight, six had also taken the opportunity to attend the third at Kyoto. An overwhelming majority, more than 50%, had attended the Third World Water Forum and a majority of public officials who are still serving belong to this group. The dominance of the Water Forum participants had definite implications for the type of response received in the survey.

Many factors account for this type of skewed participation. Due to budgetary constraints, the government is unable to send an adequate number of participants to these conferences and depend on sponsors to support such participation. Until the Second World Water Forum, the donors confined their support largely to public sector organizations. The academicians, NGO representatives, members of Civil Society and individual experts in the field were left out and had no means to participate. Again, the donor funds are generally available from the development projects that are controlled by specific ministries. While nominating participants, ministry's officials get the preference without any consideration of the necessity of involving other concerned people having relevance to the themes of the conferences. For these reasons, United Nations conferences have largely been attended by officials of the Ministry of Foreign Affairs and by people from the ministries connected directly with the achievement of the goals of these conferences (e.g. MDGs) such as the Ministry of Health, Ministry of Education, Ministry of Social Welfare and Ministry of Youth and Sports. Ministries connected with water resources management were not considered very relevant for participation in the United Nations and international conferences. Officials who had attended the United Nations and international conferences have since retired and do not maintain any active interest any more in those activities. This is the main reason as to why there is hardly any response on these conferences. The question of participation is very crucial for the success of an international event: unfortunately, this is an issue that is rarely discussed in any appropriate forum. In order for the conferences to be useful, there is a need to review the overall system of sponsorship of participants and the process of their nomination. Government is sometimes prone to look upon these sponsorships as a means of patronage rather than as important opportunities to put across its own point of view and influence relevant decisions taken in different forums.

10.2.2 Overall View

The overall view of the respondents about the megaconferences is quite positive. Measured in the scale of 0–5, the overall average is 3.28 while the average of those conferences attended by the respondents is 4.1. United Nations conferences generally and the United Nations Conference on Environment and Development in particular were graded as the best, closely followed by the Second World Water Forum. The Kyoto Forum received the lowest ranking.

10.2.3 Impacts

Response to this topic was quite diverse, though an overwhelming number (42%) thought that the conferences had at best a marginal impact on them or their institution. Of the respondents, however, 26% thought that the conferences have significantly changed the policies, programmes and projects of their institutions. These changes would not have happened if these conferences had not taken place. Of the respondents, 10% considered that these conferences were excellent. They have radically increased their knowledge-base, and have improved their working practices significantly. A similar percentage considered these conferences as a forum for meeting old friends and making new ones, though the conferences really had no lasting or visible impacts on them or their institutions. Only two respondents have crossed two alternatives each. One crossed B and C while the other crossed C and E. D and F were the two options that were not considered by any of the respondents.

If we look closely at the various categories of responses, two diametrically opposite views on the impact of the megaconferences emerge. If we combine the percentages of responses against option A (10%) and option B (26%), we have 36% of respondents who think that the conferences were excellent and they had significant impact on water sector governance and management against 42% who think that the conferences had a marginal impact. The difference in perception can be explained in terms of locus and focus of a particular respondent. Those dealing with policy issues will have different views on these conferences *vis-à-vis* their counterparts responsible for implementation. Similarly, public sector officials have their own views on many of the water sector issues highlighted in these conferences that influence their evaluation of the impact of these conferences. Each person has their own perspective on issues they consider important, and it is difficult to expect an absolutely neutral response to questions centring those issues.

10.2.4 Strengths

The respondents were asked to list a maximum of three strengths of megaconferences. Three respondents did not answer this question. The rest listed a variety of reasons which are condensed into seven items on the basis of close similarity of views expressed. The frequency of the condensed statements is then calculated from each completed questionnaire and the number noted in parenthesis. The result of this exercise is produced below.

- Provide opportunities for interaction of diverse views, opinions and visions on water governance at a global level and help create awareness and better understanding of global water issues (5).
- Provide the forum for sensitizing the policy-makers as well as the stakeholders on the importance of water resources and its utilization in the context of future needs (4).
- Create a positive impression on policy-makers and practitioners about the imperatives of Integrated Water Resources Management (IWRM) (3).
- Help create pro-environment consciousness (2).
- Provide access to new information and help fill knowledge gaps (2).
- Outcomes of such megaconferences enjoy legitimacy and credibility at a global level (1).
- Induce the politicians to make commitments to pursue social development policies in their home countries (1).

- Ministerial Declarations provide guidelines for the future direction of the water sector (1).

It will be seen from the above that the respondents consider the megaconferences very crucial in raising global awareness on burning social development issues. IWRM and environmental issues are noted as positive outcomes of such conferences.

10.2.5 Weaknesses

The response to this topic followed a similar pattern as in the previous topic. Three respondents did not answer this question while the others mentioned a wide range of reasons. Attempts were made to condense the diverse statements but it has not been possible to do that in all cases. There are some statements that are unique in character and defy their consolidation. What is presented below is a mixture of four condensed statements along with a number of individual ones with the frequency mentioned against each of them:

- The unnecessarily vast nature and format of these megaconferences make them impersonal experiences for most participants, making these as "forgettable events" (4).
- No realistic action plan for implementation of the decisions is taken and governments carry on with business as usual (3).
- They are expensive to attend compared with the outcomes (2).
- Resolutions are non-binding on the member countries (2).
- Too many disjointed topics lacking focus and depth (1).
- Presentation of papers takes the majority of the allocated time with very limited time made available for open discussion (1).
- Big and powerful countries dominate the proceedings and the developing countries hardly get any opportunity to put across their views and in cases where they do, these are not given any serious consideration (1).
- Delegates from the developing countries, including the ministers and senior government officials, do not take these conferences seriously (1).
- A number of activities seem redundant because they prove to be peripheral in nature, something like a "ritualistic fanfare" (1).
- The outcome of megaconferences is pre-determined and does not originate from the conference proceedings (1).
- The environment of megaconferences is more of festivity than of serious concerns for achieving results (1).
- The outcomes of megaconferences do not have any significant impact on national programmes (1).

The size and format of the conference are perceived by the respondents to be the most serious weakness of megaconferences. Other factors such as focus and depth of issues discussed and time available for open discussion by participants are related to the format and organization of these conferences. Also related to it is the question of cost-effectiveness of these conferences and the ability of interested persons to afford participation at their own cost. The allegation that the proceedings of the megaconferences are dominated by the big and powerful countries and the developing countries do not get any opportunity to participate adequately is a very serious indictment and needs careful design of conference programmes. The prevailing notion that the important resolutions are informally decided long before the actual opening of the conferences generates a lot of misgivings. Nobody would dispute the necessity of doing a lot of preparatory work prior to the actual start of the conference proceedings but these should be processed for finalization as an essential part of conference proceedings in an open and transparent manner.

The other issue of major concern has been the neglect of follow-up action on the decisions taken. Vision and policy statements need to be linked with a plan of action for implementation. Finance and logistics are important elements towards realizing lofty policy statements. While these conferences have been very meticulous and prompt in developing policies, they have not generally invested the time and energy to chart out an implementation programme with firm financial commitments from the donors.

10.2.6 Cost-Effectiveness

Three respondents did not answer this question. The responses of the others are listed as follows.

- In terms of expenses involved in holding these conferences and the outputs derived from them, they are not cost-effective at all (10).
- They could be made more cost-effective by implementing decisions taken in these conferences across the globe (3), by pruning certain activities in order to focus on result-oriented tangible issues, which could be more realistic, achievable, and less geared to media attention (1), by restricting participation to people who are able to contribute (1), and by arranging these meetings on a regional basis in a focused manner (1).
- These (the conferences) are games people play with some positive impact (1).
- Whatever the cost may be, megaconferences are essential to sustainable management of water resources (1).

There is an overwhelming view that the megaconferences are not cost-effective. The respondents have also raised some good suggestions for making them more cost-effective. The issue of implementation has surfaced here again and it seems that lot of people consider the success of these conferences contingent upon the relative success in implementing some of the decisions taken there.

10.2.7 Documentation and Information Dissemination

Three respondents did not respond to this question. Others, overall, have given a positive assessment. In this case, it has not been possible to condense the statements and they are reproduced here from the questionnaire to get a flavour of what the respondents have to say on this topic.

- The quality of documentation and information dissemination has always been a strength of such major water events.
- All kinds of documents are available in these megaconferences produced by different countries, international financing institutions and NGOs. The documents are useful.
- I do have some documentation that I collected during the Second and Third World Water Forum meetings. They are generally of good quality and are helpful in my professional activities (teaching and research). However, the dissemination process of the documentation is less than optimal.
- Repetitive, purely theoretical and less weightage given to important issues.
- Documents were well-prepared. But there are too many things in the documents. The documents were not well designed to suit actual implementation of programmes.
- Except for Ministerial Declaration, I do not have copies of other documents. I do not think these were widely circulated. It is also difficult to publish a single document covering all the topics discussed in the meetings. With limited access to the documents, I am not in a position to comment on their usefulness and quality.

Respondents do not have many complaints about the documentation; however, there are problems regarding packaging and dissemination.

10.2.8 Practical Results: National Policy Changes and New Initiatives Including Water Sector Reforms

The section is actually a consolidation of three topics as they appear in the questionnaire. However, national policy changes and new initiatives are important results that may originate from these conferences. In order to avoid duplication, it is better to treat them together. The respondents have also answered them in a way that makes it more convenient to consider them under one section.

The response to this set of questions has rested on personal experience and knowledge of the respondents about the developments in the social, environment and water sectors. It appears that some do not have the information of the activities at ground level in Bangladesh and they have expressed their inability to make any statement on these issues. Those directly or indirectly involved with the above developments have come out very clearly in pointing to the linkages between the decisions taken in these conferences and their impact on policy and initiatives at national government levels. These two sets of responses are shown below with their frequencies.

- Not aware of any national policy changes or any new initiatives, including water sector reforms (6).

- The following conferences had some impact on the formulation of national policies and plans:

Mar del Plata, Dublin and Rio	National Water Policy (6)
Mar del Plata, Dublin, Rio and Hague	National Water Management Plan (7)
Johannesburg	Setting the target of achieving MDGs (3)
Dublin and Rio	National Environment Policy and National Environmental Management Plan (3)

- Mar del Plata, Dublin and Rio have also influenced the revision of the Bangladesh Water Development Board Act in 2000 and the formulation of the Guidelines for Participatory Water Management 2001.

10.2.9 Changes in Investments

This is one question that has received the most negative review. Three respondents chose not to respond which is an implicit indication that nothing noticeable has happened as a result of these conferences. Seven respondents have very emphatically stated that the conference outcomes have led to no investments at all. The remaining respondents have their own interpretation of the impact of these conferences on investments. These are reproduced next from the completed questionnaire.

- Investment for the water sector has not increased or decreased as a result of these conferences. Rather, emphasis has shifted from one user to another within the sector.
- ODA support in the water sector, especially in the development of small-scale water resources, is gradually on the increase.
- Sometimes a few agenda are picked up by one or more donors (e.g. MDG by UNDP) which receives noticeable funding from that donor.
- Unfortunately, it (investment) has decreased or has slowed down. This is probably owing to the fact that donors have linked investments to adopting the policies and principles prescribed from these conferences without looking into what is immediately required for the country.

10.2.10 Key Lessons

Responses to this question indicate that an overwhelming majority of respondents do not like to write off the role of these megaconferences in improving the economies of less developed countries and their people. However, they are highly critical about the organization and process of these conferences and have some suggestions to improve their effectiveness in the future. Respondents whose views may be construed as somewhat negative have not always left the matter at that and have offered valuable comments for better organizing them. The responses are thus mostly positive with some negative comments but suggestions for improvement under both the categories. Following is a sample of some of the positive and negative comments and suggestions for their improvement with frequency of their occurrence given in parentheses.

Positive

- Provide an opportunity for interaction and exchange of ideas/knowledge among water professionals from different parts of the world, especially those from the Third World countries (3).
- Facilitate an understanding of the benefits of a holistic approach to water management (IWRM) including awareness about environment (3).
- Create opportunities for introduction of new ideas and concepts that are subsequently pursued by the donors (2).
- There are elements which promise positive outcomes (1).
- Are basically awareness-building campaigns (1).

Negative

- Have not been able to make contribution, either positive or negative (1).
- Evaporate as ephemeral events and are soon forgotten. Follow up activity is poor (1).
- Seem hardly cost-effective and the outcomes are not followed up at the country level (1).

Suggestions

- Regional conferences on regional issues might be more useful and effective (2).
- Megaconferences could be organized through consensus building at the regional meetings with a much smaller number of participants and more region/country specific recommendations (1).
- Steps could be taken to enhance cooperation and collaboration among countries required for the success of such conferences (1).
- Secure donor support for implementation of some of the vital decisions taken in these conferences (1).

10.2.11 Changes in the World of Water Owing to These Megaconferences

On this question, there is both cynicism and appreciation. However, the cynics are fewer in number than the supporters of these conferences. The following is an attempt at condensing the varied statements with frequency of their occurrences.

Positive

- In terms of raising awareness about poverty, climate change, sustainability of eco-system and a new perspective about the role of water on development and livelihood, these conferences/events played a major role throughout the world. They played major roles in shaping up the water regimes in many countries, especially where good governance is being practiced (4).
- Through these megaconferences, water has become everybody's business (2).
- The world has changed but not much (2).
- The world has changed in positive ways which would not have happened but for these conferences (1).

Negative

- The world of water would not have been much different if these megaconferences were not held (3).

Two respondents did not fully respond to this question, while one respondent has given a conditional response by saying that the world would have changed had the development partners followed up the decisions by concrete actions.

10.2.12 Overall View of the Megaconferences

The questionnaire gave four options to the respondents and each person was to indicate the ones that came closest to their view. Seven respondents have chosen one option each while another six have preferred to combine two/three options to express their views. When sorted by frequency of occurrences, the following result is obtained.

- Instead of the global megaconferences, it would be desirable to organize regional meetings, dealing with regional problems and issues, and which could be focused and impact-oriented (4).
- The concept of such global conferences is good, but the present framework for organization needs to be changed radically. The events should be more focused and output-oriented. The main criterion for success should not be the number of people who attended the conference, but the quality of the results and their impacts (2).
- The global megaconferences are useful and cost-effective. We should continue with them, but only with marginal changes (1).

Four respondents have answered by combining the options C and D while another two have combined B, C and D.

The overall response to this question clearly shows that an overwhelming number of respondents appreciate the usefulness of megaconferences at a conceptual level but they would like its organization and process to be reformed. The concepts articulated at megaconferences need to be further developed and refined at regional meetings with a solid plan of action for their implementation.

10.2.13 Ministerial Declarations

Participants of these megaconferences seem highly disinterested and marginally aware of the Ministerial Declarations. The Ministerial Conferences are mostly attended by officials of the concerned ministries. Unfortunately, most of these participants have retired and they are not generally active in such intellectual pursuits. The respondents are mostly agency specialists, academicians and representatives of the NGOs and civil society who did not have any opportunity to attend these meetings.

All 15 respondents answered this question. Of them, seven stated point blank that they had no knowledge of the Ministerial Declarations, though most of them had attended at least one of these conferences. The remaining eight had attended mostly the Third World Water Forum and a few the Second World Water Forum. Their comments on Ministerial Declarations derive mostly out of their experience in these conferences. Only two of them consider that these declarations had had any influence in policy-making in the water sector in Bangladesh. The other comments are generally negative and the following statement by a respondent captures the essence of such comments:

> The Ministerial Declarations seem to be too diluted and a watered-down version of the genuine conference aspirations or objectives. The generalized statements avoid specific and contentious issues, and are less pragmatic in nature; hence, often unimplementable. Besides, the receptivity among the policy-makers and bureaucrats to these declarations is very low.

10.2.14 Views on Other Issues not Mentioned Above

Only five respondents answered this query. Others did not respond presumably under the impression that their response to other queries had adequately covered the necessary ground. The views expressed by individual respondents are listed next.

- (a) It might be beneficial to downsize the scope and format of the megaconferences so that the participants can actively engage in a wider number of events/activities. (b) Too many parallel sessions organized at the same hour (during the conference) should be avoided in order to let the attendees participate more widely. (c) The sub-themes for the conferences could be kept within reasonable limits in order to avoid the image of the conference being disjointed, pedestrian, pedantic and rudderless.
- There is a need to stress on implementation, on changing the mind set of political and bureaucratic leadership and reduce corruption in the sector.

- Donors and lending agencies should convene first before a new megaconference. They must ask experts to present the evaluation of the earlier ones. Feedback from developing countries must be collected and considered seriously.
- (a) Megaconferences can be useful once in a while. (b) Regional conferences on specific regional issues with participation of selected experts from relevant countries to share their knowledge and experience may be more useful. (c) Outcomes of these conferences should be pursued at country level by their governments as well as by the participating NGOs, such as Water Partnerships and Water Forums.
- There should be regular follow-up of the decisions of the megaconferences through the instrument of development aid.

10.3 Conclusions

From the above analysis of the survey findings, the following important issues emerge for active consideration of the sponsors of these megaconferences to enhance their standing, credibility and effectiveness.

1. *Scope*. Megaconferences may be framed in many different ways. The following options are worth considering:

    - Megaconferences are preceded and followed by regional meetings for facilitating the preparation of the agenda for such conferences, and for developing an action plan for implementation and monitoring of decisions taken in those conferences with assured donor funding. The other option could be to start up with a megaconference for developing the concepts and outlining the broad agenda and then follow up further development of those ideas into implementable action with commitment of donor funding.
    - Maintain sectoral focus within the framework of a multisectoral approach or go for a purely sectoral approach.
    - More focus on developing concepts for raising awareness versus blending of theoretical concepts with the imperatives of their implementation in the socio-cultural milieu of particular developing countries.
    - More focus on implementation issues and less on conceptual developments.

2. *Participation*. A rigorous definition of conference objectives, scope and methodology and determination of the kind and level of participation are prerequisites for the ultimate success of these conferences in achieving lasting results. Strict criteria for selection of candidates as participants other than the ministerial delegates should be drawn up and strictly adhered to.
3. *Conference format*. Conference design needs to be reviewed in terms of its scope. The need for parallel sessions may be evaluated on the basis of objectives set for such a conference. The various themes selected for exposure and articulation must bear strict scrutiny by experts in the field and a few areas may be identified for full concentration of the delegates. The decisions taken in the

conferences must be arrived at in an open and transparent manner to remove the misgivings that these are predetermined and decided behind the scenes. Delegates must be given adequate time and opportunity for taking part in the discussions and to ensure that more time is allocated for open discussion.

4. *Ministerial Declaration.* These declarations need to be widely discussed in plenary sessions and properly disseminated among all participants.
5. *Follow-up of decisions and implementation.* Follow-up of decisions and an implementation action plan must form an integral part of megaconferences.
6. *Evaluation.* Every megaconference and regional meeting must be evaluated by a panel of neutral professionals on the basis of success criteria determined prior to the holding of such meetings. The outcome of such evaluations would be considered and reviewed in designing similar meetings in the future.

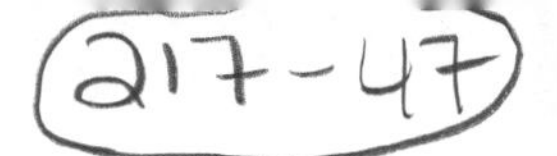

11 Megaconferences on Water: Perceptions from the Nordic Countries

Olli Varis and Terhi Renko

11.1 Introduction

The past two decades have witnessed a swarm of massive international gatherings that have all scrutinized and elaborated the future of water management and development on a global scale. Conferences such as Dublin in 1992, Rio de Janeiro also in 1992, Marrakech in 1997, the Hague in 2000, Bonn in 2001, Johannesburg in 2002 and Kyoto in 2003 have all become well-known and debated events for individuals and organizations that have their activities in the water sector, particularly those who are engaged in the development arena.

Such a congestion of important and massively participated events has, on the one hand, contributed to the awareness of water's many roles in societies and the environment in many ways. On the other hand, people increasingly ask if this frequency of international mega-gatherings with such gigantic participation is really meaningful. To give a dimension, the Kyoto meeting in 2003 attracted officially around 24,000 participants.

The aim of this report is to summarize and analyse the results of a questionnaire survey that addressed this question. The views and opinions of water experts and policy-makers from five countries in Northern Europe were analysed. These countries, Denmark, Finland, Iceland, Norway and Sweden, are called the Nordic countries in this report. Denmark, Finland and Sweden are members of the European Union.

The Nordic countries have altogether 23 million inhabitants, which accounts for 0.4% of the world's population. These countries are relatively wealthy, and their share of the global Gross National Income is 2.3%. They are also blessed with water, being able to enjoy 1.4% of the world's renewable water resources. In one sense in relation to this study, however, the Nordic countries can be considered as superpowers. That is official development assistance (ODA): this region with 0.4% of the world's population contributes 9.1% of all ODA globally. The US, for instance, with over 10-fold population is responsible of 16% of all ODA, whereas France and Germany both account for 10%, UK for 6% and Canada for 3% of that budget.

This study was carried out as part of a global-scale survey coordinated by the Third World Centre for Water Management in Mexico. This Centre performed a worldwide analysis on the topic. In addition, a series of geographically focused studies were performed including Southern Africa, Japan, India, Bangladesh, and the current study on the Nordic Countries.

11.2 The Approach

The survey questionnaire was identical to the one that was used in the global-level analysis. The following eight megaconferences were included in this survey:

1. United Nations Water Conference, Mar del Plata, 1977.
2. International Conference on Water and the Environment, Dublin, 1992.
3. United Nations Conference on Environment and Development, Rio de Janeiro, 1992.
4. First World Water Forum, Marrakech, 1997.
5. Second World Water Forum, the Hague, 2000.
6. International Conference on Freshwater, Bonn, 2001.
7. United Nations Conference on Sustainable Development, Johannesburg, 2002.
8. Third World Water Forum, Kyoto (partly also in Osaka and Shiga), 2003.

Here, they are classified into three groups: United Nations Summit Conferences (Mar del Plata, Rio de Janeiro and Johannesburg), their preparatory events (Dublin and Bonn) and World Water Forums (Marrakech, the Hague and Kyoto). In many of the subsequent analyses, this classification is used principally owing to the very different functions of the conferences.

The United Nations summits have a key policy forum role in the United Nations System. They outline the strategic directions of the United Nations activities for at least a decade ahead. Their preparatory events are expert and stakeholder gatherings for distilling the key issues to be brought to the summits. They are a part of a follow-up mechanism; in the case of Rio and Johannesburg, this is organized by annual follow-up events with the acronym CSD (Council for Sustainable Development). The theme of CSD meetings in 2004–2005 was Water, Sanitation and Human Settlements.

The third group of megaconferences are the World Water Forums (WWFs). They have not an official status in the United Nations System, even though they prepare a Ministerial Declaration. The WWFs were originally modelled after the prestigious World Economic Forums which have gained a high media attention and political prestige in the past years. However, the WWFs have evolved in quite a different direction. Their major functions have been very much expert oriented and related to information dissemination, dialogues, etc., and the Forums have grown in size enormously.

A questionnaire was sent to approximately 300 water experts and/or policy-makers in Denmark, Finland, Iceland, Norway and Sweden. The request was sent by e-mail, and a reference was indicated to a Web page, which contained a detailed description of the study, as well as a link to the downloadable questionnaire. The completed questionnaires were requested to be returned either by e-mail or by fax. Thirty-one individuals returned the questionnaire. Seventeen responses came from Sweden, eleven from Finland, one from Iceland, Norway and Denmark each.

Views from those that had attended one or more of the listed megaconferences as well as those who had not attended any of them were solicited. It was assured that no comments would be attributed to any of the respondents personally.

The questionnaire had altogether 16 questions. The questions and answers of the 31 Nordic respondents are summarized below.

11.3 Participation

This question asked about which of the listed megaconferences the respondent had participated in. There was one individual who had been to six megaconferences, one had been to five of them, and nine that had not participated in any (Figure 11.1). The remaining 16 respondents had been at one to four events. As an average, the respondents had been at 1.5 megaconferences.

There was only one individual who had been at the Mar del Plata Conference. Three had been at the Johannesburg event. A relatively low number of respondents had also been to the Rio, Marrakech and Dublin events (4 each). In contrast, the attendance to the Hague, Bonn and Kyoto had been much higher: 11, 8 and 11, respectively. No one had been either at all WWFs or at all United Nations summits.

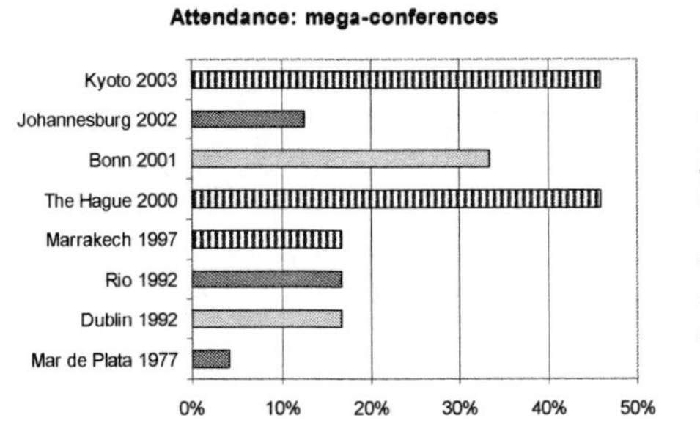

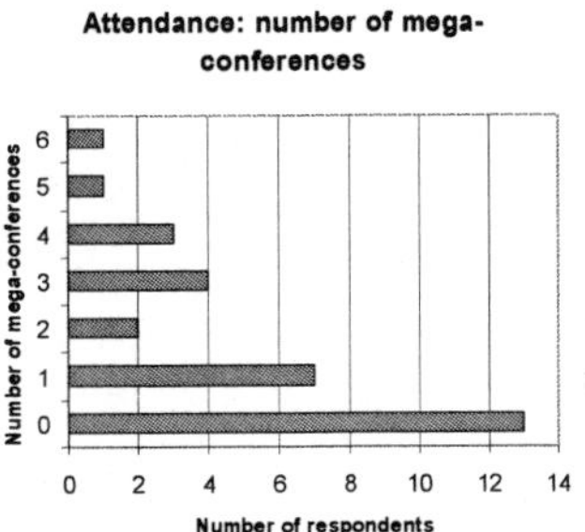

Fig. 11.1 Percentual attendance of Nordic survey respondents in the eight megaconferences. The conferences are classified into three groups: United Nations summits (Mar del Plata, Rio, Johannesburg), their preparatory events (Dublin, Bonn) and WWFs (Marrakech, the Hague, Kyoto). The number of attended meetings by individuals is also shown

11.4 Overall Views on Each Megaconference

This question scanned the overall view of the respondents on each of the megaconferences under survey:

> Based on your current knowledge of these megaconferences, please state your overall views on each of the event(s) in a scale of 0 (very poor) to 5 (absolutely excellent), based on your perception of their outputs and impacts. Use 3 for average. If you have no specific knowledge of a conference, please say N/A.

The average view of the respondents was relatively positive (Figure 11.2). With the scale used, the overall average was 3.30. The average of those conferences that the responder had attended was somewhat higher, at 3.49.

Interestingly, the results show a decreasing trend of appreciation. The most ancient one, Mar del Plata, was graded the best (even though only one respondent of this study had attended), whereas many of the attended ones were graded low. There is one factor that explains a good part of this trend. The first two megaconferences attended received almost invariably a good grade, from 3 to 5, with only one exception (Figure 11.3). The respondents were far less happy with the third and still less satisfied with the fourth or fifth conference they had attended.

The Hague was graded highest of the WWFs. The other Forums were seen as the poorest among all the megaconferences under survey. The United Nations summits were graded higher than their preparatory events.

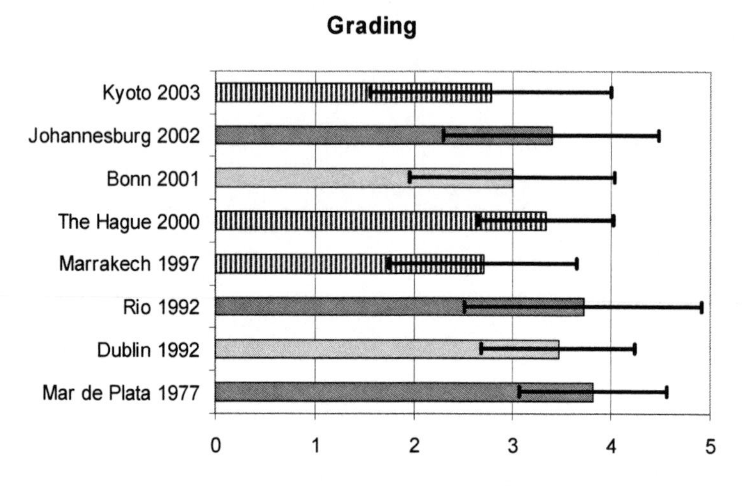

Fig. 11.2 Grade (1 = very poor, 5 = absolutely excellent) given by the respondents of the eight megaconferences. The average and standard deviation are shown. The conferences are classified into three groups: United Nations summits (Mar del Plata, Rio and Johannesburg), their preparatory events (Dublin and Bonn) and WWFs (Marrakech, the Hague and Kyoto)

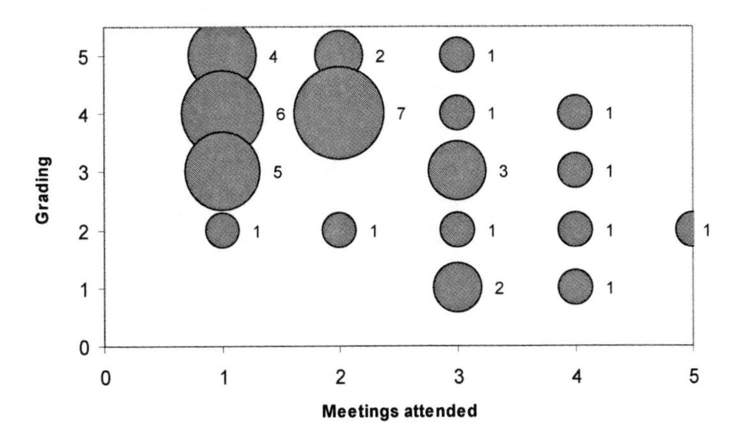

Fig. 11.3 Number of grading given (1 = very poor, 5 = absolutely excellent) as a function of the number of meetings attended. The two first megaconferences attended received high grades but the gradings decreased sharply after that. To analyse this figure, e.g., seven of the respondents gave a grade of 4 to the second megaconference they had attended

11.5 Impacts

The impacts of megaconferences on various levels, individual, institutional, managerial, etc., were investigated with the following question:

> Impacts of megaconferences: Irrespective of whether you participated or not in these megaconferences, please select which one of the following comments most closely reflects your overall views on all the megaconferences as a whole.
>
> a. The conferences were excellent. They have radically increased my knowledge-base, and have improved significantly my working practices.
> b. The conferences have significantly changed the policies, programmes and projects of my institution. These changes would not have happened if these conferences had not taken place.
> c. The conferences had at best a marginal impact on me and/or my institution.
> d. The conferences had no perceptible impact on me and/or my institution.
> e. It was pleasant to attend the conference(s), meet old friends and make new ones, but the conferences really had no lasting or visible impacts on me or my institution.
> f. These were mostly forgettable events. For all practical purposes, it would not have mattered much whether these events had ever been held or not. They simply did not leave any footprints on water management.

In addition, there was an option to present additional, verbal comments on this matter.

The highest number of respondents chose option B, meaning that the megaconferences had had a significant impact. Altogether, 12 had indicated this alternative. In addition, five chose B with one or more other options. An overall score was calculated (see Figure 11.4) so that if there was only one letter indicated, the corresponding option had a score of 1; if two options were indicated, each chosen option received the score 0.5, etc. The sum of these scores is the number of respondents, in this case 28.

Besides B (13.83), option C got a high score of 9.33. The other options were scored 2 or slightly less. The most negative option, F, was not chosen by any of the respondents.

There were a few additional comments. One respondent agreed only partially to the "political" outcomes of the events. One mentioned that the events have produced or marketed ideas and slogans. One responded that his/her personal views were more positive than those of his/her institute. And one respondent's institute, a United Nations agency, takes its policy guidance from some of these conferences.

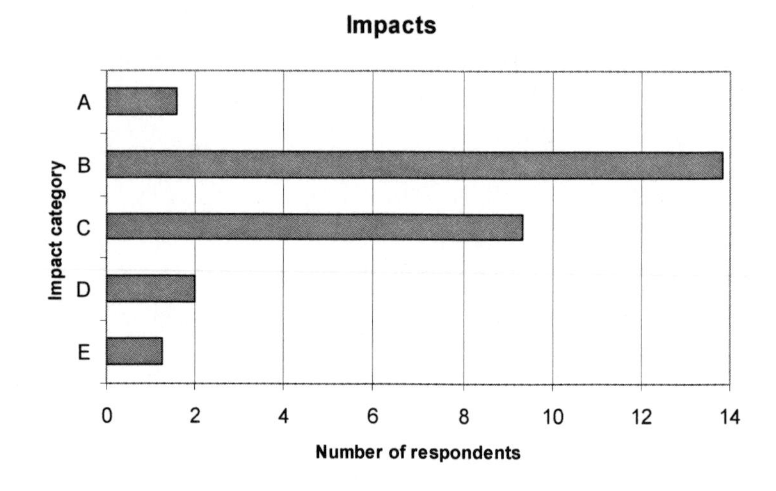

Fig. 11.4 Number of respondents judging the impacts of megaconferences to each of the five classes (from A to E) as defined in the questionnaire

11.6 Strengths

This questionnaire surveyed respondents' perceptions with regard to the strengths of megaconferences. A maximum of three reasons was asked for.

The response was very diverse. Twenty-six respondents expressed an opinion on this matter. They are condensed to the following 14 points. Several references to individual countries, statements and external events were made but they have been removed. The number in parenthesis shows the number of respondents who mentioned this point:

- Dialogue, exchange of experience between people with different backgrounds, learning and networking (18).
- Publicity to water issues and raising public awareness (10).
- Stakeholders: Empowerment, participation and communication (7).
- Influence politics and policies, particularly in the long term (7).
- Finding consensus and shared views (7).
- Raising political awareness to water issues (5).
- Bringing politicians closer to experts (3).
- Possibility to lobby and contact funding institutions (2).
- Initiating and launching new initiatives (3).
- Global approach and outreach (3).
- Contribute to open information exchange (1).
- Strengthen the water sector (1).

One of the replies was very different from the others:

1. Mar del Plata: international start-up activity; ambitious preparatory reports; national activities initiated. Follow-up: Drinking Water Supply and Sanitation Decade.
2. Dublin: Last-minute rescue activity to get water addressed in an international perspective; without it water would only have been addressed in the poor way in which it was at Rio. Follow-up (GWP and WWC): Dublin principles widely disseminated and referred to.
3. Johannesburg: Water Dome activities (effort to highlight water properly; opportunity to disseminate statements).
4. The Hague: Water Vision process involved ambitious preparations; regionally inspiring; strong follow-up in three international dialogues – financing, climate and food/environment.

11.7 Weaknesses

In analogy to the previous question, the weaknesses of megaconferences were surveyed by a question that asked the respondents to list the maximum of three most important weaknesses of the events. The points raised are listed below. The number in parenthesis indicates the number of respondents who mentioned each point:

- Too diffuse, too many interests, unfocused, confusing (12).
- Discussion is too general, not enough practical implications (10).

- The dual role of policy and expert gatherings creates plenty of mess. The need for political consensus is seen as negative. In particular, there should be a clear distinction of United Nations policy-related events (such as Rio, Johannesburg...) and open WWFs (8).
- Too many, too frequent, too repetitive megaconferences (4).
- "Hijacked" by lobbies, some politicians, certain scientists and multinationals (4).
- Too expensive compared with outcome (4).
- Far too large (3).
- Dissatisfaction with ministerial statements and agreements (2).
- Exaggerated expectations, the event itself becomes more important than its outcome (2).
- Developing country interests are too poorly represented (1).
- Used by some organizations, politicians, etc., to wash their reputation without any policy changes (1).
- Not a good tool for raising public awareness (1).
- No opinion (4).

In addition, there were three comments that did not easily fall in the above categories.

- Rio: Nothing new on water, fragmented; Dublin: voting procedure was a failure; Johannesburg: conceptually poor in addressing of water issues; Bonn: limited, conventional; Marrakech: just starting up activity for WWC, statement was no real international statement; Kyoto: inconsistent, Ministerial Declaration process was a disaster, no use of its results.
- Water is basically a human right recognized among many by the United Nations System.
- No space for true dialogue and learning; too much preaching to the converted; too many environment and water ministers instead of the finance, defence or health ministers.

Not surprisingly, the events were seen to be too large, too messy, too unstructured, too diffuse and so forth by about one-half of the respondents. One-third saw the dual role of policy-making and expert gatherings confusing in one way or another. They were also seen as been hijacked by various lobbies and interest groups. A number of other noteworthy points were listed as the above bullets indicate.

11.8 Cost-Effectiveness

The cost-effectiveness was addressed in the questionnaire with the following itemized question:

The megaconferences are often expensive to organize, and the costs seem to have increased significantly in recent years. For example, the costs of organizing the United Nations Water Conference in Mar del Plata, or the First World Water Forum in Marrakech were modest. The cost of organizing the Second World Water Forum was much higher. The cost of the Third World Water Forum was very significantly higher than the Hague Forum. The cost of the Secretariats alone for the Forums are quite high: normally well over $10 million.
Based on your perception of their outputs and impacts, what is your view of the cost-effectiveness of these events?

The cost-effectiveness was already addressed by four respondents in the context of the previous question. The following were the items mentioned:

- They are cost-effective (1).
- Not very cost-effective (5).
- Not/not at all cost-effective (4).
- Difficult to compare, impossible to judge or no answer (5).
- Follow-up is crucial – good follow-up could yield considerable efficiency benefits (2).
- Costs are high but the benefits are also high, manifold and not only monetary (4).
- Costs are high and benefits are very difficult to measure owing to scattered and long-term benefits (1).
- Cost effectiveness is not a proper yardstick for megaconferences (1).
- They are too large and unfocused to be efficient (3).
- They are too frequent to be efficient (1).
- Costs are high, but international conflicts cost much more: there is always a premium on cooperation. But rising costs are a concern, so megaconferences should not be held too often (1).
- Particularly the Third WWF of Japan (Kyoto) in 2003 evoked criticism in this respect.
- It would be better to give the funds used for running the Secretariats as direct aid to support public water utilities in poor countries.
- The attendance fee to Kyoto was too high for many people from developing countries to participate. Besides, maybe some national developed country delegations could have been smaller and thus part of the money spent for participation fees could have been spent with the actual water programmes and projects.
- The Kyoto figures are alarming, and not in proportion to the outputs. The Dublin meeting in 1992 must have been one of the most cost-effective ones, taking the forceful Dublin Principles into account.
- It is not necessarily that simple – a high cost would be acceptable if there are results in return, but Kyoto was an example of too much bureaucracy...which did not help in creating results; quite the opposite.
- Three venues at the Kyoto Forum were one hour or more in travelling distance from each other: one venue would have been enough, probably also one-third the number of participants.

It would be very interesting to know, particularly with regard to the last comment, who should have been excluded from the Third WWF? Or how, from a following comment, the "important stakeholders" should be defined and how the participation should be controlled?

- The costs should be cut down to the Hague level in the following ways: controlling the amount of participants but including important stakeholders; the facilities and accommodation (especially those paid by organizers) could be more modest; paper load and material costs should be cut.

Moreover, there were several responses that would rather have belonged to different questions:

- I hope they had impact on other people. Not anyone I know though.
- They are too political.

11.9 Documents and Information Dissemination

The documentation of the megaconferences was the topic of the next question:

> Do you have adequate documentation (reports, papers, proceeding, etc.) from any of these conferences? If so, which ones? How useful has this documentation been? Overall, what are your views on the quality of documents you have seen, and information dissemination processes of these megaconferences?

The documentation of these events has consisted of a spectrum of products with varying goals. Official policy documents, background documents, workshop and seminar briefs, brochures, books, etc. The supply has been escalating in recent years and the use of the Internet as a distribution channel has changed this field very much.

The respondents did not structure their replies very much in accordance to what was said above. Some gave very general comments such as those listed below:

- Ignorance (none, maybe) (4).
- Satisfaction (OK, adequate, useful…) (8).
- Too much material (1).
- Quality varies very much, some is very useful (5).
- Information dissemination is very good within, but not outside, the water sector or developing countries (3).
- Documentation is highly useful and well available particularly through the Internet (2).
- Have some reports, do not use them very much (1).
- The wide spectre of choice among such events has probably attracted many participants (1).

Some respondents specified the following.

- Mar del Plata: first broad documentation in the water field; Rio Agenda 21 widespread, not very useful for water; Johannesburg: action plan poorly disseminated; Dublin Principles widely disseminated and used; WWFs: no reports/no useful reports.
- Documentation tended to be mostly United Nations-type negotiated documents.
- The Hague and Kyoto: very good documentation and dissemination.
- Dublin Statement, Bonn and Kyoto: Dublin Statement has been useful; value of conference documentation has been more as reference than improving my knowledge base; lot of side event material, which however has not been very useful for me personally (lack of time to read and review).
- To improve the dissemination processes I could propose the following. (a) The major outputs should be written on short leaflets with clear pictures and/or graphs, be printed in all major languages and be distributed to all members of parliaments all over the world. This will cost some money but, taking into account the total costs of megaconferences, I suppose not too much in view of the results the action might bring. (b) The major subjects that are discussed could be refined to a game or a discussion forum or any other kind of form that the younger generation is interested in. These games, etc., should be playable on the Internet during and after the conference. Schools, colleges and universities could then take a benefit of utilizing the internet to participate in the conference, also by commenting on the discussions held.
- The Bonn documentation, report and the Bonn keys have been useful, the Dublin Declaration/Principles, as well. I keep the Johannesburg Plan of Implementation in my briefcase, and quote a paragraph or two every once in a while in my presentations; the Kyoto documentation was laughable (but characteristic of the conference) with the exception of the Camdessus Report.
- Rio Conference was a hallmark, with Agenda 21 and the conventions, and Johannesburg with its action plans. These outputs guide our work; it is less clear with the WWFs, but basically the analytical reports and action plans have all been rather useful; the dissemination efficacy varies from conference to conference. The WWFs have been much more narrowly targeting only water professionals.
- The recent 5 years have been completely different from the past in this respect owing to the Web. Particularly the Hague Forum had excellent Web pages. The Kyoto Forum was less successful in this regard. This aspect is crucial. The materials distributed at events have grown very much and become diverse. These events are important sources of materials and publications. The official policy documents, in turn, have become perhaps less important, particularly with regard to events such as WWFs.

Quite diverse views have been given. After all, the discussion should be divided more clearly between official policy documents, general background reports and informative brochures. Typically, those who target their comments to political documents do not hold the WWFs in high esteem, but have a more positive view on United Nations summits and their preparatory events. This is obvious. Accordingly, those who seek materials for academic purposes, getting updated views on what different actors are doing etc. appreciate the way in which information distribution at the WWFs has evolved.

Many had a critical view to the broad and ample information availability at the megaconferences. In our view, this is fairly strange because it would be very questionable to restrict the different interest groups of distributing their materials openly. Particularly at WWFs, the ample supply is one of the key starting points of a successful open event, and the participants must rely on their own skill in navigating through them.

11.10 Practical Results

After the question on documentation comes the issue of practical results with the following question:

> In your view, did any of these megaconferences have yielded positive, implementable and lasting results? If so, please give examples from the specific megaconferences at regional, national and/or global events:

Again, the response was quite diverse and included the whole spectrum of opinions and views. The most typically addressed issues were as follows:

- United Nations summits unify and create international rules for different actors and formulate financing mechanisms for implementing them (4).
- Creating public and political awareness (4).

The other views were more diverse and specifically articulated as follows:

- Has brought up the importance and complexity of water issues; little impact outside the sector itself.
- Mar del Plata, Dublin and Rio have partly developed our thinking related to water. Such thinking (an integrated approach to water management, stakeholder participation, etc.) has probably permeated into national policy- and decision-making. If this is the result of a specific conference, or more a general trend that would have taken place anyway, is hard to say. Implementable actions on the ground, probably not. Hopefully such conferences would have a positive impact on resource allocation, but OECD statistics do not really indicate this. Similarly, there is no real proof that water is higher on the national political agenda today – a good example is the "lack" of the water issues in the Policy Reduction Strategy Papers.
- It was important that the Bonn Conference expressed opinion against international loans.
- Sometimes these resolutions may have biased interpretations.
- When applying funding for some water projects, it is very useful to refer to these events, which even the politicians tend to know. Knowledge of the water issues has been raised higher by these conferences. It is another question of how much reflection there is on the ground level water projects.

- These types of results are more dependent on the successful pre-negotiations rather than on the event itself.
- Overall change of policy trends. Difficult to give examples, though discussion on shared waters is one.
- Most of them have had some positive results although not very visible in the short term.
- From a local government point of view, the succession of these conferences has enabled us to make policy-makers and donors aware of the role of local government, and of the need for decentralization in securing access to fresh water and sanitation, and the need for local water governance; as an example, the Camdessus Report has highlighted at a high advocacy level the need for financing at the local level (sub-sovereign finance); both the African Water Fund and the EU Water Facility will make funding accessible for the local level; the Bonn Conference made it very clear that privatization of water utilities is not the only possible solution in providing equal access to fresh water and sanitation.
- Mar del Plata, Rio and Johannesburg have all been first-grade milestones in international policy-making on water. They have influenced greatly the policy agendas for a decade ahead in each case. Dublin and Bonn were preparatory events for Rio and Johannesburg, respectively, and crucial in this regard. WWFs have had their merits in information dissemination and as open platforms of discussion.
- Rio was very special with a broad impact, based on Brundtland Report, excellent preparatory work and extraordinary good Chairman and Secretary-General. Furthermore the very special international situation just after the end of the cold war helped to create a sense of opportunity. The conventions signed in Rio (Climate, Biodiversity) or just after (Desertification) also helped to strengthen the impact.

The key political actions and documents were recognized and highlighted by some respondents. However, since many listed them under the question on New Initiatives, these answers are listed in that context. Six respondents expressed no view on this matter.

11.11 National Policy Changes

Thereafter, there was a question on how the megaconferences had changed national policies:

> Are you aware of any policy change in your country which would not have occurred without one or more of these megaconferences? If so, which policy or policies were changed because of these events?

The responses to this question were on average much shorter than those to the previous questions. Again, six responders had either no view or responded with 'no'. The three most frequent points that were mentioned were (specific country names are replaced with "my country" or corresponding expression):

- Not much in my country. But they have had an impact on the bilateral and multi-lateral commitments and/or development cooperation policies (9).
- Difficult to specify but they are key references in policy discussion and definitely influence policies (4).
- It has influenced national policies through the EU Water Framework Directive and EU Water Initiative (3).

Four participants highlighted Agenda 21 of the Rio Conference as having influenced particularly local environmental policies nationally. The other points mentioned were as follows:

- It is difficult to tell what would have happened without megaconferences, in any case, I do not see where else the dialogue could have happened.
- Growing awareness of the importance of IWRM and demand management has had impacts on the content of many aid projects, stressing, e.g., institutional cooperation and water legislation. Especially Dublin and the Hague contributed.
- My country's water policy is in line with Johannesburg.
- After Dublin the development agency of my country openly expressed the policy that water is an economic good and commodity, thereby adapting to the corporate agenda. This Spring the agency has presented a Strategy for Water Supply and Sanitation, which does not mention water as a human right, which has been criticized by the United Nations Commission on Human Rights. Instead the strategy says that it does not matter if the cat is black or white, the only thing is to deliver water, no matter who does it. This is also the actual policy of the EU, which is the home of the main global water companies.
- The national policies have, on large issues, taken advantage of the forum and political visibility these megaconferences have given. In particular, the Johannesburg Conference.

11.12 Changes in Investments

The impacts on investments were surveyed next with this question:

> Has the investments availability for the water sector in your country increased or decreased by these conferences, which would not have occurred unless these conferences had taken place? If so, please provide the direction and a rough estimate of these changes. Or have these events had no perceptible impacts on investment availability in your country:

Seventeen participants did not respond, or responded with 'no' or 'not much impact'; one responded with little impact, and one had the view that they had influenced markedly. One referred to an earlier question. The other points mentioned were as follows:

- The water and sanitation sector of my country is basically very sound and effective, and to the great majority under public regime. It has only to a little degree been indirectly affected by megaconferences.
- At this time the EU Water Facility is the only financial mechanism dedicated solely to the water sector.
- Our government seems to add development cooperation funds, water being among the key sectors.
- This is probably the really sad part of the story. OECD figures indicate no substantial increase after 1989 (especially not grants, some increase to the mid-1990s in loans). Actually, from the end of the 1990s there is a decrease. My feeling is that there has been an increased interest in my country – but I have no figures to support this. There is also a strong focus on the water and sanitation sector – and not so much on IWRM. I am not sure the conferences have had any impact on this.
- Unfortunately, there is much more talking than real action.
- Still waiting for the finalization of the modalities of the EU Water Facility and of the African Development Bank's Water Fund.
- Probably no significant change. Investment level is already quite high and sufficient.
- The Rio Conference had perhaps a negative role on the funding of the water sector owing to a low recognition of water issues. The situation shows some change resulting from Johannesburg and other events.
- I am unable to give figures but water is now higher on my country's development agency's agenda.
- No considerable impact can be observed in the water sector as a direct result of the findings coming from these megaconferences. Investments, new initiatives.
- Difficult to judge, but our corporate sector has actively followed the conferences.

11.13 New Initiatives Including Water Sector Reforms

This question surveyed the views on new initiatives launched as a result of megaconferences:

> In your view, did some new initiatives originate from these events, which otherwise would not have occurred? What are these initiatives? Also, did these events contribute to water sector reforms in your country? If so, in which areas?

One group of responses specified certain new initiatives to certain megaconferences.

- Mar del Plata: Water Decade (4).
- Dublin: Dublin Principles (7); the "corporate agenda" of Private-Public Partnership (1); development cooperation policy in the water sector (1).
- Rio: Agenda 21 and other sustainable development activities (5).
- The various dialogue processes (1).

- Marrakech: Establishment of World Water Council (1).
- From Marrakech to the Hague: The World Water Vision (1).
- The Hague: Dialogues (2); World Water Visions (1).
- Johannesburg: Millennium Development Goals (5); IWRM (4); EU was greatly inspired to take the EU Water Initiative and later the EU Water Facility (2); WSSD Action plan (1).
- Kyoto: Camdessus Report (1); Water Action Inventory (1).

The above list includes some points from the Practical Results question. Another group approached this question in a more general level.

- There are more initiatives impacting on developing countries than Nordic countries (4).
- Perhaps the EU Water Framework Directive proceeded faster in the 1990s as a result of raised awareness (2).
- Regional networking and partnerships have partly emerged from the megaconferences (1).
- Perhaps increased attention on groundwater resources (1).
- Hopefully more focus on water governance first, infrastructure investment afterwards (1).
- Owing to the vast dimension of events such as Kyoto and Johannesburg, there are innumerable initiatives at various levels. Some of them fail, some not. Without bringing people together links would obviously be weaker (1).
- Increased awareness of water issues and increased efforts for dissemination of information and cooperation. Increased visibility internationally (1).

Altogether, 15 respondents gave no answer to this or mentioned that they are not well aware of any such initiatives. Surprisingly few participants, after all, were able to specify even the key initiatives listed above.

11.14 Key Lessons

The key lessons from the megaconferences were surveyed with the following question.

> What in your view are the key lessons (positive and negative) that we can learn from these megaconferences?

The responses given to this question can be classified as positive, discursive and negative lessons. The positive lessons mentioned were as follows:

- Essentially, they have facilitated and broadened the scope of discussion (3).
- They have unified views and brought policies to international agenda (2).
- Particularly important in raising public awareness (2).
- They are central and necessary components in the new multilateral diplomacy and policy-making for sustainable development (2).

- They can be organized and they have effect (1).
- They have brought the really huge challenges to discussion (1).
- They have produced international agreements (1).

The negative key lessons mentioned are outlined below:

- They should be more focused; better coherence with other than just the water aspects; closer to politics (7).
- They are inefficient particularly in terms of resource use and implementation (2).
- There is too little historical continuity in the context of megaconferences (1).
- Instead of ideological yes and no discussions, we should have policy-oriented research-based results on real experiences (1).
- They should be better prepared, particularly in a scientific context (1).
- The slogans promoted are too general as universal solutions (e.g. economic good, appropriate and low-cost technology, community participation, gender, private sector, PPP) although the world is very diverse (1).
- There is a risk that megaconferences are used for manipulating policy-making by lobbies (1).
- Unrealistic goals, lack of commitment, conflicts among different priorities (1).

The discursive response was more diverse and more difficult to be condensed into short bullet points than the positive and negative one. Many of them provide the following recommendations.

- It is crucial to bring different stakeholders to the same table ("multi-stakeholder dialogues"), and not just for decoration (Kyoto) but for real debate (Bonn and Johannesburg) (1).
- Their various functions (political, policy-level, social, linking, multi-stakeholder dialogues, dissemination, etc.) should be made clearer. Prioritizations would be needed to make the profiles of different events less indistinct. In WWFs, Kyoto in particular, this has yielded a mess (1).
- To have lasting results, they should contribute a financial commitment or have access to one.
- It is wrong to look at these conferences out of context, the Monterrey Conference needs to be mentioned as well as the Millennium Declaration of 2000 to make sense of the outcomes of WSSD.
- Process tends to become more important than substance. It is important to keep pressure on political systems – but ministerial sessions take far too much time and effort compared with what they generate. How will a conference make a contribution (value added)? How will it address the recipient of the results?
- Big events are necessary for keeping the world informed and promoting awareness of water problems and challenges. Without any large-scale promotion, water issues risk to lose their political significance. The scale of the future conferences should be more modest and some money should be spent instead to practical work. From talk to actions was the objective that the Third WWF was promoting, and which should be emphasized, but we still talk more than

act. Still, the focus of megaconferences is necessarily different from that of some smaller events. In my opinion, the purpose of the mega-events is to raise awareness, media attention and discussion, not really the practical planning and implementation. Megaconferences should be followed by smaller events for practical implementation of the initiatives.
- Seven participants gave no reply, three replied that they were not aware of key lessons, and two referred to one or more of the previous responses in this context.

11.15 Changes in the World of Water

This question was a very general one:

> In your view, would the world of water be any different if these conferences had not taken place? If you think the world has changed, in which ways has it changed?

The response was more diverse than in the case of the previous questions. Consequently, condensing the material was not as easy as above, and therefore the following list is fairly long, but extremely interesting and informative.

- There is new thinking now for a sustainable use of global water. They are also important for broadening the mind. Solutions seldom originate from one idea but from the interaction of many ideas and experiences.
- There has not been much to offer for developing countries. The megaconferences have made it easier and more legitimate for Western water companies to penetrate the developing world. Poor countries have become increasingly under the grip of Western capital. The only way to development is to respect national sovereignty, meaning that poor countries should be assisted to develop their national institutions on their own terms. In this way they can develop and keep their national competence. A good way to start is real disarmament and allocation of funds to the water sector. Massive debt relief should also be carried through.
- The megaconferences have led to frames of operation in the field of water resources management.
- They did add to the discourse and changed the prevailing paradigms.
- Connecting water internationally to related fields.
- Hopefully it has promoted its importance through media. I am not so sure to what extent it has promoted seriously science-based knowledge. On the other contrary, we have several new journals and many of the older ones have expanded their scope.
- The sustainability and IWRM issues are better understood as well as the holistic approach; access, subsidies and governance still open.
- I think the world would have changed the way it is changing anyway. However, if we think that the world is getting more globalized (and this is something we can argue about, not least "who" gets more globalized), then I think it is essential

that the "water sector" in its widest sense also needs to follow the development (to handle increased complexity, for instance). Thus, I do not think these conferences have changed the world – I think they are a result of the demands of the changing world.

- There would be less knowledge and understanding about the water issues and thus probably less funding or other support available. Some argue that the money should instead have been used working at ground level and less spent on the megaconferences. But I hardly believe that an equivalent amount of money would have been spent on practical water projects instead, if the conferences were not organized. The allocation of funding is still mostly a political issue and the politics just are not that simple.
- Every event will change the world in its part – more or less. Global awareness has increased and knowledge-base in developing countries has grown. Networking has become more effective.
- More questions are asked about privatization: it is not all black and white any more, and no one dares to say any more that privatization is the only efficient alternative; there is more understanding in the need to build governance and management capacity at the local level.
- Hard to attribute the changes to the megaconferences, *per se*. However, the Rio event and, to a lesser degree, Johannesburg event, certainly focused attention on environment and sustainable development. CBD and UNFCCC, as well as CCD, are positive developments.
- Increasing openness, multidisciplinarity and interconnectedness of water discussions, both within the water sector and between various sectors, is extremely important. The role of various forums has become very different owing to the spread of e-mail and the Internet. Getting personal contacts, broad and quick views of what is happening, and similar aspects have grown in importance because detailed documents, correspondence, etc. follow very different modalities than they did 10 years ago. Massive international events obviously address this need at least partially.
- These megaconferences have contributed to activate governments to act in the right direction. However, bad governance and politics have always been present in the process, which may have not lead to any significant (positive) change.
- Policy documents and agreements have strengthened small stakeholders struggling with water issues.
- Hopefully fresh water will become more integrated into other sectors including agriculture.
- Important awareness-raising functions and forums for technical discussions mean that new action at national and local levels has been possible.
- They have had very significant impacts, and raised the water issues to the highest level of discussion within the international community. The implementation is slow, but improving.

Seven participants did not reply, three replied that they were not aware of changes in the world, and four referred to one or more of the previous responses.

11.16 Overall View

The overall view of respondents was approached in the following way:

What is your overall view of these megaconferences? Which of the following statements come closest to your view?

a. The global megaconferences are useful and cost-effective. We should continue with them, but with only marginal changes.
b. The conferences have now become one big "water fair", with a lot of activities but without much thought as to their relevance, appropriateness, outputs or impacts. There is no coordination between events, no clear focus, and their cost-effectiveness leaves much to be desired.
c. The concept of such global conferences is good, but the present framework for organization needs to be changed radically. The events should be more focused and output-oriented. The main criteria for success should not be the number of people who attended the conference, but the quality of the results and their impacts.
d. Instead of the global megaconferences it would be desirable to organize regional meetings, dealing with regional problems and issues, and which could be focused and impact-oriented.

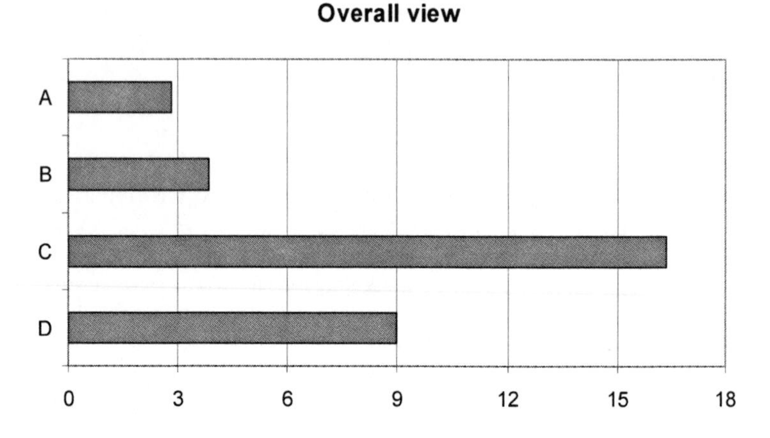

Fig. 11.5 Scores of the overall view to megaconferences. The options A, B, C and D are specified in the text. A is the most positive and D the most negative option

Altogether 28 participants out of 31 responded to this question. Many combined two or three grades. If two grades were given, the corresponding classes in Figure 11.5 received a score 0.5 from this respondent and, if three grades were specified, the score was 0.33. With this method, the alternative C scored highest, 16.33, implying that over half of the respondents were positive to the concept of megaconferences but had strong reservations concerning the way they have been organized. Option D, with the preference to impact-oriented regional conferences, scored 9. The most positive options, A and B, received the lowest scores of 2.83 and 3.83, respectively.

11.17 Ministerial Declarations

The most political part of the WWFs, the Ministerial Declarations, was also included in the questionnaire with the following question:

> At many of these conferences, there were Ministerial Declarations which have relevance to the water sector. Please give your opinions on the following questions:
> - Are you and your colleagues aware of these Ministerial Declarations? If so, please identify the conference whose declaration you consider the best and had most impacts.
> - What are your views on the relevance, appropriateness and usefulness of such Ministerial Declarations?
> - Has the water policy and/or priorities of water programmes of your country been affected by these Ministerial Declarations? If so, please briefly provide examples.

As was the case in the replies given to some of the earlier questions, the outcome in this case was very diverse. The individual responses were too different to condense into a few main points. Therefore, they are listed as such, after some editing. They are grouped into three groups as defined below.

1. *Positive*

- Mar del Plata; Rio Agenda 21; National opportunity to act: Mar del Plata, Johannesburg, Dublin; GWP followed after Dublin; SIWI and Swedish Water House after the Hague.
- Of course! We do our best to influence them! Bonn and Bonn Keys and Johannesburg (also the informal declaration); the drafting process reveals where the problems are (i.e. in the final declaration the problems are not necessarily visible any more); the drafting process of the declarations is the most political part of the conferences, the place for "twisting arms", for forming positions and building coalitions – ridiculing the political declarations (and I know that many people would like to, and find it more intelligent to do so) is not understanding the process behind them, and its value; it is important to have the declarations, because they always offer at least one line or two that can be later quoted and referred to ("in XY, ministers agreed so and so, hence..."); the linkages cannot be seen that directly. The Bonn Conference (along with Johannesburg and the MDGs) secured that water issues are on the top of global agendas and more funding is being directed into implementing them.
- Rio and the Hague; useful but the potential for improving such declarations has now been exhausted for some years. The Kyoto Declaration was not a big step forward from the Johannesburg one a year earlier.
- Probably the Hague declaration was the most influential. It definitely was the best disseminated out of the post-Rio Water Conferences. They are appropriate and useful if they are focused and have an action plan. Action plans should have clear targets, timetables and budgets, and assign responsibilities.
- The declarations from 2000 and 2002 have been useful for setting up common goals.
- No doubt Dublin. The Rio declaration was not about water declaration but important. Johannesburg was an important follow-up to the Hague.
- Yes to all.

2. *Positive and negative aspects, often scepticism for practical significance*

- In my country, very few water colleagues read Ministerial Declarations, and they are seldom highlighted in the press or at universities. Nevertheless, among the establishment, they are referred to as powerful instruments. The main thing is whose interest stands behind these declarations.
- Some impact at national level but no impact on the international level.
- The Johannesburg Summit affects my work daily where the sanitation was added to the MDGs. It has an impact, but how far is too difficult to evaluate. The Ministerial Declarations are very difficult to comply with. Most often they are signed without reflecting on reality.
- General, not very specific.
- Dublin: good and bad results (lowest appropriate level vs mere economic good). How about involving the heads of states? Have they not done this in some other sectors? Should be explored.
- The Dublin Statement was important. The later declarations have formulated the same thing in different words, but I see little development. They have not been very useful, because they are not binding. They are more relevant to developing countries. I do not know whether any policy has been affected.
- I am aware of this. I think that the Ministerial Declaration from the Hague is quite good, but that it has had minor impact in reality. Not very important, no changes in my country.
- I am aware to some extent, but have not really bothered to study the declarations in detail. In principle it is important to commit the high-level decision-makers and politicians. In practice it unfortunately may often happen that ministers tend to forget the whole issue after the conference?
- Rio and Johannesburg have been very important. Similarly, Dublin and Bonn have been important as their preparatory events. A doubt arises as to whether WWFs are appropriate places for Ministerial Declarations. Many of the merits of these Forums are inflated in public discussion owing to weak ministerial statements.
- These are necessary, but they have limited impact. However, since they are negotiated, they carry long-time weight. In some cases, such as the Rio Declaration, the impact is very important. However, in many cases, it seems that the effort of negotiation does not correspond to the limited effect. I cannot state with certainty that water policy in my country has been affected by these declarations.

3. *Negative*

- I don't know about the Ministerial Declarations, and my own country certainly was not much affected. But I am aware that other statements made in these fora, declarations made, agreements signed and so forth, have been instrumental.
- I am aware, but do not believe they are the main outcome of the conferences, e.g., the Kyoto Ministerial Declaration did not include many things that had an actual impact. In the case of the other conferences, the relevant information concerning them is usually something other than the Ministerial Declaration.

- I feel that the declarations are not that important. To improve the awareness of politicians on water issues could bring better and more sustainable results. An ordinary politician feels him/herself important when proposing actions to be taken rather than acting according to some declaration given somewhere far away.
- Good for newspapers.

Four respondents gave a 'no' answer, 5 not aware, no impact or a similar statement, one referred to an earlier question.

It is noteworthy that those who argued rigorously were more often positive than negative. A majority of the respondents considered at least some of the Ministerial Declarations significant but followed with an expression of scepticism to the practical importance.

11.18 Other Aspects and Issues

Finally, an option was given to the participants of the survey to bring forward any other matters that they did not mention in the context of the previous questions:

> Please give your views on any other aspect(s) and issue(s) of global megaconferences not mentioned above. This could be as long as you wish.

There were several lengthy answers to this question, although only 11 participants responded to this question. The outcome was highly informative. A summary of the responses is as follows.

- Five points were mentioned in this response:

1. The difficulty to link research and policy hampers the impact of megaconferences. We need to find forms to bridge them and I do not think that megaconferences are a good way to do it! Whose responsibility is it to take research into policy? This should be the goal of regional meetings that I think would be more efficient than megaconferences. Such meetings or, preferably, series of meetings are needed to really have an impact on policy. It should be a platform to produce policy-relevant input that can synthesize information on natural variability, human factors, climatic change, etc.
2. The need for consensus means that only symbolic agreements can be found, and the impact is very limited, even if agreed to by many. But maybe just the fact that there is an agreement is more important than the real impact?
3. There is Western Europe/United States domination in megaconferences. Therefore, local initiatives should be preferred now. Policy culture differs globally; different mechanisms are needed to channel scientific knowledge in different parts of the world.
4. The importance of public awareness as a facilitator of action has often been neglected: no matter how many megaconferences are arranged, declarations signed, statements written, etc., if public awareness, which makes policies acceptable, is missing (this is also necessary in order to make politicians dare to

take action, otherwise they will not be elected). This simple truth has been neglected. Megaconferences are too isolated; local initiatives would have a larger impact.

5. In spite of the above, we need global meetings. But they need to be more focused and be based on sharing experiences from regional, ongoing series of workshops, etc.

- Three conference categories with different aims, strengths and weaknesses – (1) United Nations System conferences like Mar del Plata, Rio, Johannesburg; (2) international with limited invitations like Dublin, Bonn; (3) non-governmental like WWFs – have been most useful through:
  - mobilization of resources and preparatory reports which would not exist otherwise;
  - projects generated;
  - way of activating governments which can benefit from political goals and Minister Declarations if they like to, in order to nationally defend certain activities;
  - follow-up most effective if taken care of by some organization such as GWP, GEF of family or international organizations (Dialogues); other conferences that have not been mentioned, similar to Bonn, are Harare and Paris (no megaconferences though).

- This reply listed six items:

1. The hypothesis underlying the above questions seems to be that these conferences are a waste of time, which this survey has set out to prove. Questions 3 to 15 are loaded in that direction, bringing me to doubt the scientific value of the whole exercise. I find its approach simplistic.
2. It is certainly important to always keep in mind the cost-benefit equation of these megaconferences, and we always need to ask whether another conference is really necessary. It is also reasonable to ask whether there have been too many water-related conferences. But all of those analysed did not deal with water only. It is wrong to look at them out of context: Johannesburg was part of a chain of other events and conferences not referred to here.
3. Such conferences serve as international benchmarks and they should not be underestimated, e.g. the conventions on climate and biodiversity were both launched at Rio. Does that mean that they would not have happened in the absence of Rio? A hypothetical question that can never have a definite answer. Are we to dismiss those two conventions as useless? Well, for all their undeniable shortcomings they are the kind of instruments that our generation of mankind has been able to come up with, which at least to my mind does not make them useless.
4. Their role as benchmarks also means that they serve as venues for related initiatives to be launched. An example is the EU Water Initiative launched at Johannesburg and later followed by the EU Water Facility. The same hypothetical question may be asked here: Would the EUWI have happened if WSSD had not taken place? Maybe and maybe not, we shall never know.

5. Taking a critical look at the international conference machinery is certainly legitimate. But a scientifically correct approach should look much more in detail at what each of the past conferences has achieved. Each document or convention adopted should be analysed in detail. Their benefits can only be assessed after a considerable time. Perhaps that time period should be about 10–15 years. It would now be a good time to evaluate the outcome of UNCED in 1992.
6. Sometimes the benefits are intangible and still important, such as the declaration resulting from the Dublin Conference (also in 1992). How do you quantify the benefits of general ideas that later came to impact on much of the current thinking related to water?

- There has been a sequence of slightly different types of conferences within the water sector. The overlaps, specific issues or the steps forward or development from one to next is not very much thought out.
- I think they serve a purpose; at least, they have done this. I am not too sure about how they could and should be arranged in the future. The crucial test is if there are institutions and mechanisms that could take care of the implementation of the decisions that have been taken at megaconferences. If these efforts fail, it is regrettable, but it would not be fair to blame these events as such. I see them and other conferences as meeting places for ideas, experiences, for pledges, etc. And their *raison d'être* is not primarily to be scientific meetings, but more platforms for high-level policy-making. But they are not implementing agencies. That is not their role. The link to implementing agencies is much more expensive and more difficult as compared with arranging megaconferences.
- It should be revised why these megaconferences were initially organized. Only then, the message of their key objectives could be reported properly. A mega-event cannot directly contribute much on the ground level work, and it is misleading to claim that it would be able to do so. It is better to clarify the purpose of different kinds of events. Both are needed in some rational scale: megaconferences targeting on global- and national-level awareness raising and policies, and smaller planning events for practical orientation.
- Global megaconferences can be useful if they are output-oriented and come with concrete action plans. Otherwise their impact remains largely academic. Unfortunately, I believe that Kyoto was an example of this. Dublin largely failed because it was organized too shortly before Rio and outside the official preparatory processes. Consequently, it had no influence on the Rio processes: an opportunity missed.
- Policy documents are important but as important are small-scale initiatives focusing on local needs. Developing country decision-makers and scientists must be more involved. The industrialized countries have to listen to the needs of the developing countries and collaboration projects must be developed projects, which are truly implemented and sustainable.
- My opinion is that discussions and recommendations seldom move beyond jargon even at practitioner-oriented conferences. This could be overcome with

more professional and innovative discussion moderation. Another general opinion is, in spite of jargon-like recommendations and maybe poorly executed presentations, that they are important in information dissemination and as meeting points between professionals. Maybe their main contribution is to be a meeting point between founders, decision-makers, professionals and practitioners? That it might be the only place where all these mingle? Other conferences might not be considered important enough to draw donors and decision-makers. However, it might be that the coffee breaks are more interesting than the actual sessions.
- I am a fervent believer in multilateral cooperation, and I feel privileged to have had the opportunity of participating in many of these conferences. In concluding, I wish to quote Maurice Strong who ended the Rio Conference by saying that the whole process had been an extraordinary human experience. I fully share that view.

11.19 Challenges

The outcome of this questionnaire survey yielded a rich set of information on the various perceptions of 31 water experts from Denmark, Finland, Iceland, Norway and Sweden on the most important, recent water-related megaconferences. The following set of five challenges can be distilled from this information.

- How to handle documentation and openness with expanding participation.
- How to deal with the needs and aspirations between global and regional level, as well as the sectorial and multidisciplinary dimension.
- How to manage the very diverging awareness level of water experts.
- Motivations to attend: range from meeting with people to political ambitions.
- Politics vs science needs clarification.

11.19.1 How to Handle Expanding Participation, Documentation and Openness?

All the possible international agendas and recommendations call for participatory policy-making processes. It is also clear that, at the global level, more and more people have the interest and possibility to attend important events such as the open megaconferences that have been analysed in this study. With the spread of education and democracy, the number of people that have a theoretical possibility to attend such meetings must have grown manifold since the organization of the Mar del Plata Conference in 1977.

The Nordic countries, however, belong to the minority of the world's countries that have not changed much in these two respects. Their education level, number of water experts, political system, economic wealth, etc. have not undergone major changes since 1977. Therefore, it is not astonishing to note that many of the

Nordic participants feel very puzzled when facing the huge crowds of people that these days attend international gatherings, particularly in Asia (Figure 11.6).

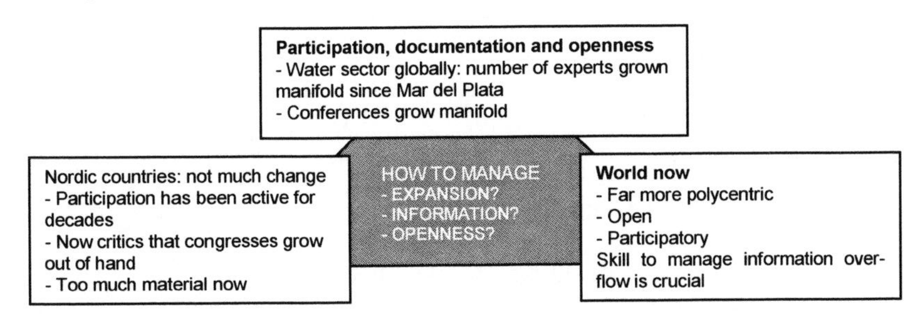

Fig. 11.6 Challenge 1: Immense growth of number of professional, openness and information flow

Kyoto's Third WWF was particularly criticized. However, the Nordic participants may not properly realize how small our countries indeed are, in terms of population, particularly if seen from the Asian perspective, and do not appreciate that they have themselves been privileged to send the largest delegations to many international meetings. Their total population of 23 million counts for only 0.4% of the world's population. The Nordic participation to megaconferences has always been exceptionally active. If a notable participation is aimed at in a region such as Pacific Asia (with a population of two orders of magnitude higher than in Nordic countries), it is clear that we must be ready to deal with far more massive numbers of people than in Northern Europe. Many of the countries in the region, moreover, have very recently become politically tolerant enough to allow other than formal official delegations to attend these types of gatherings. Additionally, the financial possibilities to experts and stakeholders in this region have improved equally recently to allow participation.

This is one matter that must be kept in mind when criticizing the Kyoto Forum. It does not downgrade the other weaknesses of the Forum, which are partly owing to the proximity of the Johannesburg Conference. Perhaps the Ministerial Declaration should have been left out of Kyoto in order to clarify its role as an open Forum.

It was also somewhat strange to learn that several Nordic experts would have liked to restrict the material that has been available at the megaconferences. The fact is that for many people, the active possibility of having access to such a richness of materials as has been the case in the two most recent WWFs, in the Hague and in Kyoto, has been a very important asset and opportunity. Hardly any Nordic responder appreciated the growing openness and participation of different people to these Forums. It does not need to go back more than 15 years in time to recall that in those days, half of Europe was not allowed to participate in an open international dialogue. The same situation was prevalent in a big part of Asia. Under these circumstances, it is not surprising that the number and size of global gatherings has grown dramatically, but it is somewhat strange to read that so many

Nordic experts would put restrictions to documentation and participation of these sorts of meetings.

The world has simply become more open and polycentric. The information flow is enormous and the skill to pick up the most useful information grows rapidly in importance.

11.19.2 How to Deal with the Needs and Aspirations Between Global vs Regional Level and Sectorial vs Multidisciplinary Dimension?

The megaconferences under study can be classified in many ways. There were two features that arose frequently in the responses of the Nordic experts. They were (Figure 11.7):

- Should the megaconferences continue to be global or should they become more regional?
- Should they attempt to be multisectorial or have a narrower scope?

A sizeable share of opinions was to the direction away from global gatherings and away from multidisciplinarity. Two questions arise: are there not enough narrowly focused gatherings, and are there not enough regional meetings? We feel that the answer to these both is 'yes'. However, global, multidisciplinary forums may be necessary to develop international policies.

Another question is whether there is space to develop these megaconferences. Certainly there is, and the fact that each of the conferences analysed has been unique in many ways shows that the concept evolves with time.

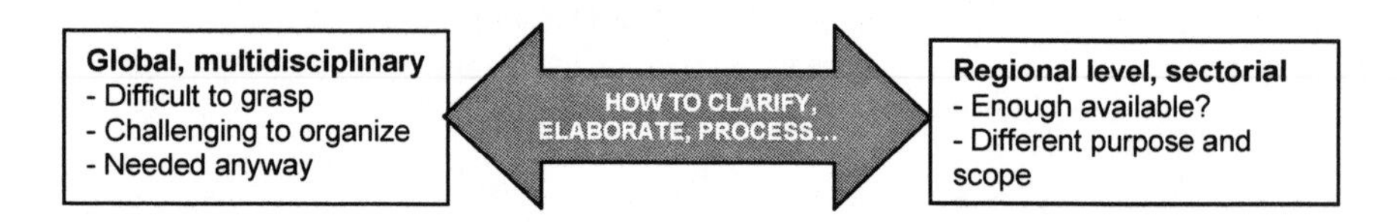

Fig. 11.7 Challenge 2: Strong contrasts between the desires and needs to approach water issues by either global/multidisciplinary or regional/sectorial way

11.19.3 How to Manage the Very Diverging Awareness Level of Water Experts?

The awareness on key issues related to megaconferences varied widely among the responded experts. Many had a thorough knowledge on the various official functions of the events whereas some did not seem to have too much of an idea of these (Figure 11.8). This broad range of insight was reflected throughout the survey. There were well-argumented, solid views on the matters surveyed, but there is a risk that they are masked to some extent by many somewhat superficial responses that in many cases did not even reply to the question asked. However, we did not want to treat the participants to the survey unequally, and wanted to include their views appropriately with the exception of omission of certain issues that were not relevant. Also, we cleaned out references to issues such as projects, events, institutions, etc. that could reveal the participants' identities, since from the outset, the survey was based on anonymous response.

Whereas awareness of water was mentioned as one of the key points in the context of several questions by many responders, it turned out that many water experts themselves were not much aware of why these megaconferences were organized and what the major outcomes were. Hence, more awareness-raising is needed.

11.19.4 Motivations to Attend: Ranging from Meeting with People to Political Ambitions

One factor that pulls a mounting number of people to these huge "jamborees", as one respondent put it, is obviously the fact that the working environment in science and administration has changed drastically in the past 15 years, owing to the Internet and other advances in information technology. It has become simple to run projects, write papers, organize meetings, etc. by communicating through e-mail, the Internet, Web conferences and other such modalities. Equally, the access to reports, project descriptions, and other documentation has soared for the same reasons.

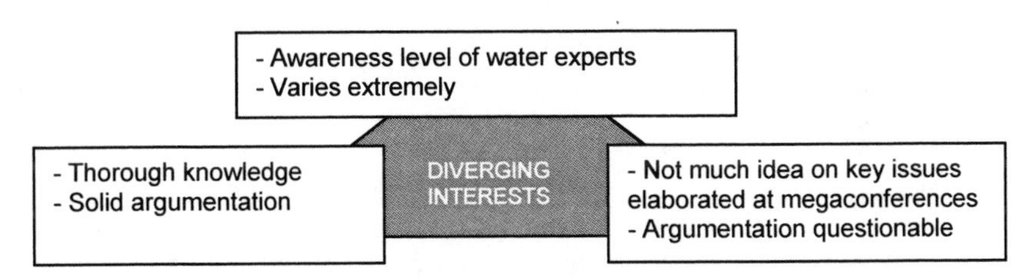

Fig. 11.8 Challenge 3: Huge diversity among water experts in interest level and awareness of key goals, modalities and outputs of the megaconferences

Obviously for many, these "jamborees" are important opportunities to meet with people whom they otherwise might never meet but with whom they might even have had very close cooperation. Surprisingly, these obvious aspect was not mentioned in this survey, but we consider it to be very important (Figure 11.9).

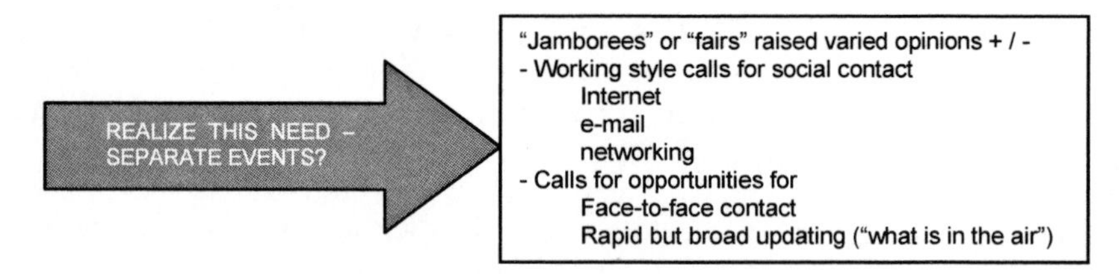

Fig. 11.9 Challenge 4: The social dimension grows in importance along with expanding telecommunication facilities including e-mail, internet, networking, etc.

11.19.5 Politics vs Science Needs Clarification

One more axis that puzzled many Nordic experts was the mix of politics and science. Obviously, many experts would be happier if the political dimension was put aside in megaconferences. However, it is a very important achievement of the past few decades that global-level policies are outlined in a setting, which is at least to some level open to experts and other stakeholders.

Many scientists would be happier with scientific conferences where they can simply conclude that more research is needed. But let us just quote the late Minister of Environment of Sweden, Ms Birgitta Dahl, who raised this issue in her opening address to the International Conference of Climate and Water in Helsinki in 1989: "We are all aware that in the society of research there are, and there should be, doubts about the absolute truth. But we as politicians cannot await the final results. Incomplete results are often used as an excuse not to take necessary measures. Too often such performances have proven to be mistakes." It is desirable that politicians and scientists discuss major issues together, rather than separately.

The official role of the meetings is thus mixed with other ambitions and aspects far more than what the case was a few decades ago. One challenge to the organizers of megaconferences in the future is to clarify the different functions of such events and particularly make a clear distinction between governmental-policy function and more general insight-oriented expert functions. One example of this is the approach by United Nations HABITAT, which now organizes its World Urban Forum every second year as a policy forum and every other year as an expert forum. This division keeps the two functions clear but the problem then is that the intermingling of policy-makers, experts and other stakeholders is reduced.

Megaconferences are used for producing shortlists of general policy recommendations, such as the Dublin Principles, Bonn Keys etc. Whereas these may be useful outcomes of the meetings, we think that the way they are arrived at should be more carefully planned. Clearly, megaconferences have many other roles, which may even be more important than the official part. However, particularly the WWFs are often judged only on the basis of the documentation they produce, while the other merits are forgotten (Figure 11.10).

After all, the last section entitled "Other Aspects and Issues" extracted perhaps by far the most interesting outcome. It is diverse enough that we will not pick up any specific key points that we have not mentioned in these conclusions. Rather, we recommend going back to them and read them with care.

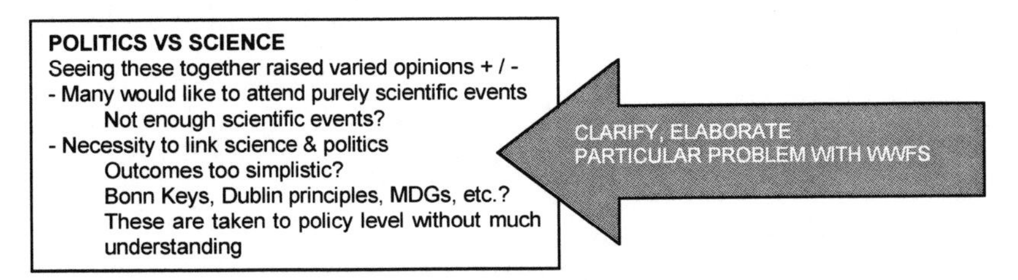

Fig. 11.10 Challenge 5: Many attendees to megaconferences would rather be away from politics and see scientific presentations. At the same time, policy recommendations tend to be simplistic

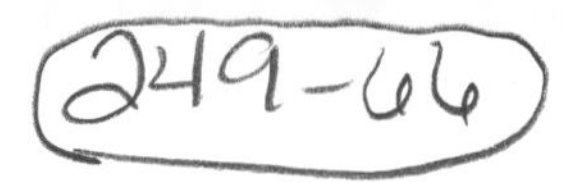

12 Megaconferences: View from Southern Africa

Anthony Turton, Anton Earle and Mikael Drackner

12.1 Introduction

As part of a global survey to evaluate the impacts of megaconferences in the water sector conducted by the Third World Centre for Water Management in Mexico, the African Water Issues Research Unit was commissioned to undertake a regional survey of the Southern African region.[1] The overarching objective of this study is to evaluate whether or not global megaconferences, often highly criticized and expensive by their nature, do have a marked effect upon local realities within the Water Sector, how they are perceived by the regional water community and what we can do to improve their impact and standing globally. This report highlights the most interesting things that came out of the questionnaires, and does therefore not treat all the subjects asked about in the actual study. It is to be seen as a contribution to the debate about the nature and future of megaconferences based on empirical research. As such it is intended to spark further discussion, rather than provide conclusive answers.

12.2 Methodology

The Southern African leg of the global survey was conducted through the sending out of a questionnaire. This was an adapted version of the original which was deemed to better fit the task at hand in an African context (see Appendix A). This was sent to approximately 200 individuals from the region as well as a number of organizations. A number of reminders were sent to try and encourage maximum cooperation and effort. Of these, 30 persons responded, putting the response rate at 15%. The highest response rates/sent questionnaires came from Botswana (31%) and South Africa (28%) which together accounts for nearly half of the responses received. The questionnaire incorporated both strictly quantitative and open-ended qualitative questions to ensure maximum output. Respondents that

---

[1] In this chapter the concept of megaconferences refers specifically to the following events: United Nations Water Conference in Mar del Plata (1977), Dublin Conference (1992), United Nations Conference on Environment and Development in Rio de Janeiro (1992), Bonn Conference (2001), Johannesburg Summit (2002), and three World Water Forums (Marrakech 1977, the Hague 2000 and Japan 2003).

have not specifically indicated that their comments could be attributed to them are quoted as "Anonymous".

12.2.1 Limitations Specific to the Southern African Survey

Logistical constraints and ill-developed IT communications in many parts of the region poses a severe constraint to the amount of data that can be collected using the given approach. Poor connections, long download times, or limited time on the Internet within the region make people less prepared to answer this type of questionnaire. The small sample size makes it impossible to draw any statistical conclusions of value. This chapter should be seen to have solicited Southern African views on the megaconferences, as they happened in the past as well as which path they should take into the future. With this objective in mind the following presentation will be presented qualitatively, letting the voices of the respondents speak for themselves as much as possible.

12.3 Findings

12.3.1 Attendance and Overall View

The number of informants that have attended each conference is fairly low (Fig. 12.1), with the prime exception of the World Summit in Johannesburg in 2002. Two other conferences which stand out are the Second World Water Forum, the Hague, in 2000 and the Third World Water Forum, held in Japan in 2003. As is shown in Fig. 12.2 most of the respondents have only attended one or none of the conferences and only three have attended three or more. Subsequently, most conferences score rather high, with only the Third World Water Forum scoring less than three (indicated as moderate). However, the values indicated for the less attended conferences are thus only partly based upon inputs from the very few respondents that were there. Their high value is boosted by the opinions of people who know of these conferences from their documentation, their impact on the water sector or through other second-hand information. The only conference that had a high attendance by regional water professionals was the World Summit on Sustainable Development (WSSD) in Johannesburg, and the overall value of this is primarily based on first-hand experience (Fig 12.3).

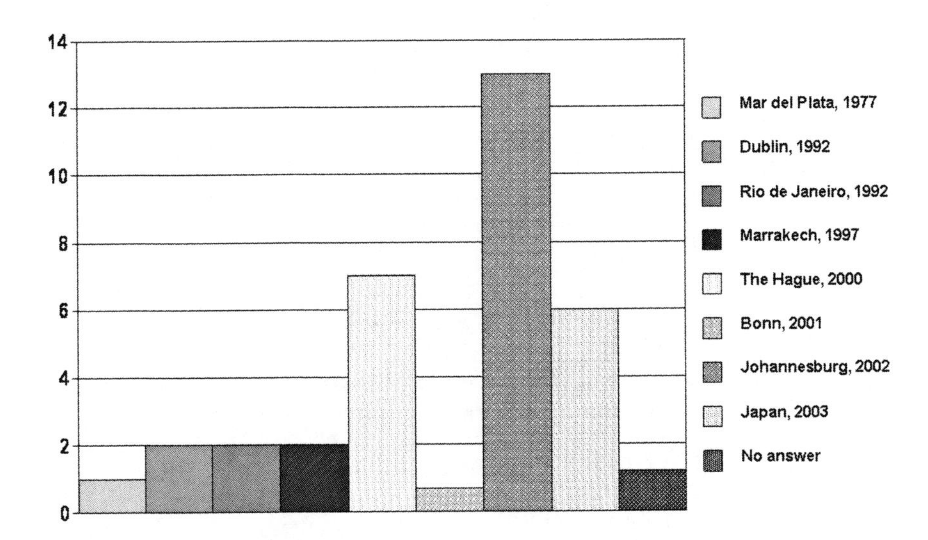

Fig. 12.1 Number of respondents attending each conference

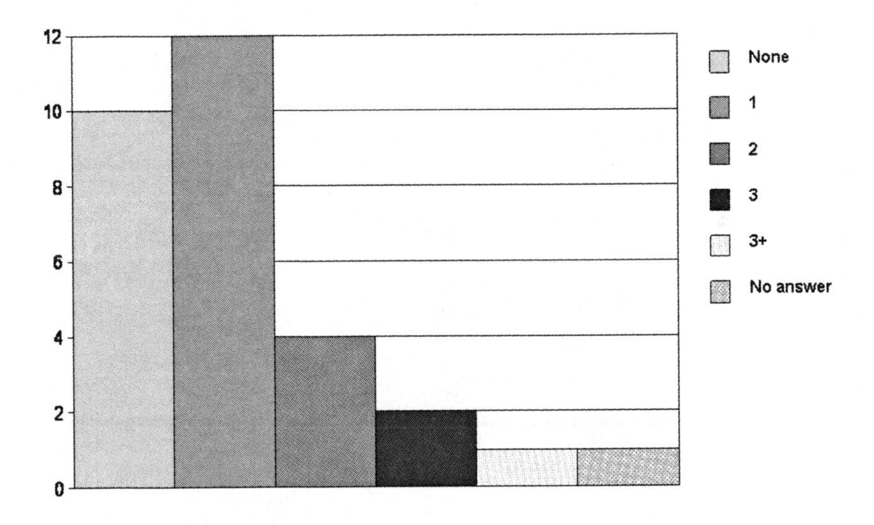

Fig. 12.2 Number of conferences attended by respondents

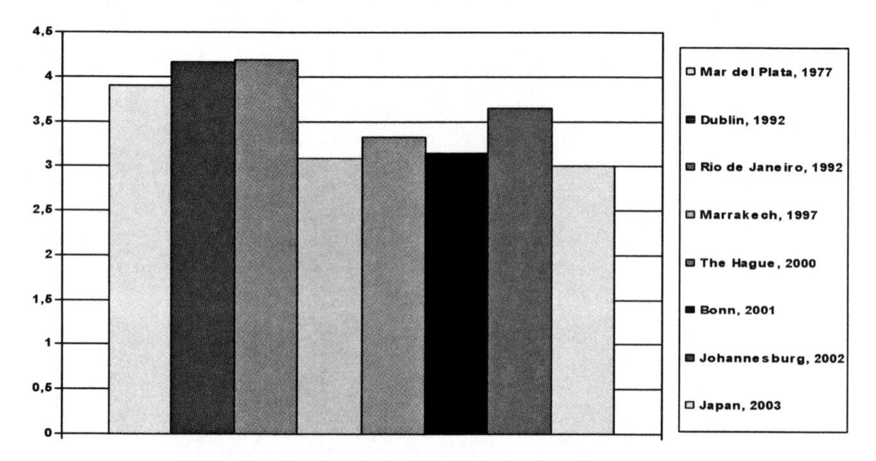

Fig. 12.3 Overall views of each megaconference

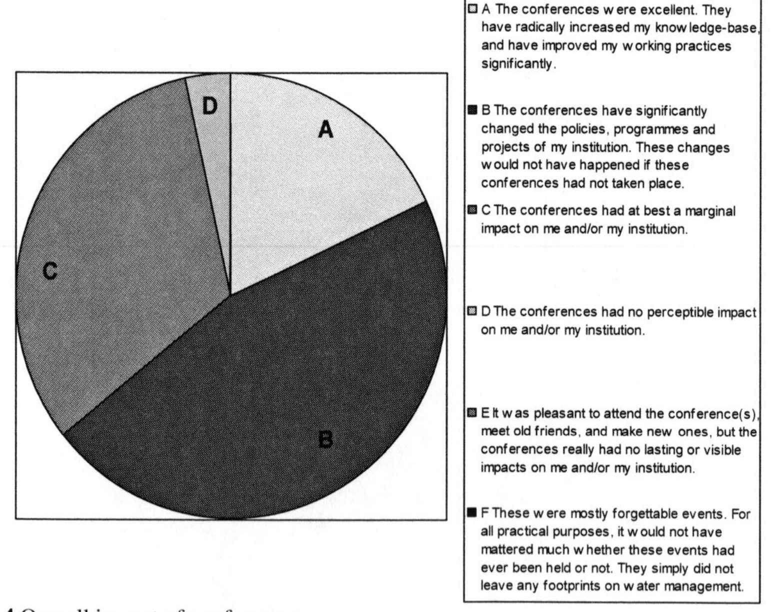

Fig. 12.4 Overall impact of conferences

When it comes to the perceived overall impact of megaconferences (Fig. 12.4 and Table 12.1), the largest portion of respondents seemed to agree with Statement B – that a significant impact on programmes, policies and projects of the respondents' institutions had been made; followed by C – a marginal impact only. However, broken down according to number of conferences attended, we see that the category that only attended one is also the most sceptical towards the impact that megaconferences have had on their own institutions. Given the small sample size these trends can be said to be indicative at best.

Table 12.1 Breakdown of overall impact according to number of conferences attended (Percentage)

Overall impact of conferences according to number of conferences attended*							
Conferences attended	A	B	C	D	E	F	N/A
0	20	20	10	–	–	–	50
1	–	40	50	–	–	–	10
2	16.5	33	33	16.5	–	–	–
2+	25	50	25	–	–	–	–

* Some respondents marked two statements as relevant to their case.

12.3.2 Strengths and Weaknesses

> The megaconference endorses ideas and thoughts and is a powerful catalyst for changing company focus and influencing policy. (Lyn Archer, Umgeni Water, South Africa)

The most important strengths of megaconferences, according to the respondents, are definitely the force they can muster to bring about global research agendas and policies and their power to foster new paradigms. As such ideas that get a wide spread stay in circulation to influence policy all over the globe for a long time. From the answers gathered it seems like there is a perception that political commitment is more easily brought about in the spotlight of the world, which can then be used to hold politicians and decision-makers accountable in a local context. One example:

> Important politicians from the countries we work in make statements to improve e.g. sanitation in their countries, which we can later use in advocacy work, holding them accountable to the content of their speeches. (Dorcas Pratt, Water Aid, Madagascar)

Although these are reportedly the most important strengths, other opinions stress the opportunity to network, the exposure to new ideas, the development of "contextual overlaps … [getting] different world views to connect" (Dirk Roux, CSIR, South Africa) which provides "almost a 'spiritual boost' in allowing cross cultural networking" (Lyn Archer, Umgeni Water, South Africa). There is no doubt that the respondents see the *ideal* conferences as opportunities to increase the knowledge base, raise the profile of the water sector or raise public awareness of the critical issues at hand. Such strengths are often the ideal perceptions of what a megaconference should be, mean or address.

In all practicality as the following section will show this is not always the case, given the views of the Southern African region. One commentator expressed the major practical benefit in the following way:

> I guess the local conference and hospitality industry in any country which hosts one of these mega-talk shops is likely to get a fairly substantial injection of cash, and that may continue to make these ghastly events attractive. (Anonymous)

In fact comments about the weaknesses of such big conferences frequently mention problems that can best be classified as logistical. Maybe the most common view is that these events have grown way out of proportion and are too ambitious to the extent where it is no longer meaningful attending. Specifically mentioned by several respondents are the parallel sessions, often spread over different venues in a large geographical area, that sometimes make it impossible to attend the sessions called for.

Other logistical constraints mentioned in the responses were language barriers and the cost of attending the conferences.

Regarding outputs, there is a strong opinion that these tend to be watered down and generalized as "one size fits all thinking" (Maria Amakali, Department of Water Affairs, Namibia), in an attempt to find a common denominator. Furthermore, resolutions taken or decisions made seldom take into account the lack of capacity and subsequent implementation problems that poorer countries wrestle with. In fact, there was some concern expressed that the attendees from developing countries were mostly politicians and very few sector specialist which at the bottom end would have to implement the resolutions agreed upon on the ground. One commentator felt that civil society and communities had little more than a spectator role.

There is also a strong feeling that the conferences are being used as proxies for furthering the agendas of various interest groups or countries. The credibility of the events is seriously damaged by practices such as described in the following example:

> [Weaknesses include] countries pushing [their] own agendas – in fact I have heard that typically the outcome of a megaconference is lobbied and decided before the conference is held. (Anonymous)

There is a feeling that conferences constitute little more than a "tradeshow for richer countries" where the developing countries have relatively little say. A common view in general proved to be that the mega events are very much a "talk-shop", which boasts few tangible results.

12.3.3 Cost-Effectiveness

Unfortunately the sample size is too small to say anything conclusive about the perceived cost-effectiveness of large-scale events. It is clear however that the direct costs for organizing a megaconference and the related costs incurred on the participants are two totally different issues.

> Direct costs for organizing are unavoidable but cost for participation is not easily justifiable compared to outcomes of the conferences. (Anonymous)

Similarly another commentator expresses her feelings about the costs involved attending a major event:

> They make it impossible for the people we are talking about to bring reasonable representation. (Anonymous)

It has to be recognized that the allocation of scarce resources in the Southern Africa region for the attendance of a conference that might not deliver anything concrete is not justifiable. While Southern Africa and other developing regions need to expand their influence and attendance at these conferences, the high costs of participating makes it a hard objective to achieve. To address this issue, several respondents call for a better focus, which would allow for more concrete outputs and better value for the individual attendee.

12.3.4 Practical Results, New Initiatives, and Policy Changes Due to Megaconferences

One of the serious questions we must ask ourselves is to what degree the mainly theoretical outputs of megaconferences can be implemented. To what degree do they inform policy and what are the new initiatives coming out of them? In the Southern African context, one of the biggest criticisms relates to the relatively small practical implications major events are perceived to have. When asked about such impacts of megaconferences *in general,* respondents answered in the following way (Table 12.2).

Table 12.2 Breakdown of respondents that answered favourably to the impacts of megaconferences on respective areas according to number of conferences attended (percentages)

	Practical results	New initiatives	Policy change
Overall	69	71	61.5
0	71	71	57
1–2	53	59	65
2+	100	66	33

However, it is very hard to take such a general stance, and more discrimination is needed.

> Positive results do come. After the Rio de Janeiro and the Dublin conferences there is more environmental awareness within many countries and various countries have modelled their water laws in accordance with IWRM principles. (Anonymous)

Some of the conferences stand out as revolutionary, while others quickly slipped into oblivion. Notably the Dublin, Rio and Johannesburg conferences stand out as examples of events that have had an impact locally, either on actual results, policy, or thinking. Of the World Water Forums the Hague meeting was the most recognized, while Marrakech received no mentioning. Kyoto got few favourable comments.

> I would argue that the period between 1977 and 1997 (including United Nations Convention on Non-navigational Uses of Inland Waters) radically altered the way water as a resource has been conceived. The law of diminishing returns, however, seems to have set in at the moment. Perhaps the problem is that many perceive the World Water Forums as state/business oriented meetings that have little to do with the real needs of real people. (Anonymous)

Shifts in direction are however not always due to megaconferences. An example:

> The results vary very much from conference to conference. (…) Rio Conference consolidated the views that were for decades being debated and as it took place, after the end of the cold war countries discussed the environmental issues in a more global level without fears. It is very important to acknowledge the importance that the end of Cold War had in the water agenda. (Anonymous)

In the case of South Africa several respondents have stressed that the changing policy in the water sector is less due to the direct impact of ideas promulgated at the megaconferences, and more due to internal dynamics created by the country's transformation process.

Although most respondents seem to agree that megaconferences do create "buzzes", highlight new ideas and raise the profile of environmental issues in general, there is greater hesitation and frustration with the question of whether or not the ideas trickle down to the ground in Southern Africa. Here are two examples:

> I am aware of several policy changes as a result of the conferences – but whether these policies are/can be implemented is another question. (Anonymous)

> I do not have a very high opinion of the usefulness of megaconferences in general, nor do I believe that they affect political decision-making very significantly in most countries (there are a few exceptions, where the citizens are interested and informed and push their leaders to follow through on statements and commitments made at some megaconference or other platform). Their impact on actual operations on the ground is minimal in my experience. (Anonymous)

12.3.5 Ministerial Declarations

The Ministerial Declarations are most definitely known to the majority of respondents (71%), with the Johannesburg Declaration not surprisingly being selected as the best. This is not to say that it actually was the best. Rather it points to the fact that this is the Ministerial Declaration that respondents are most likely to be very familiar with, many from first-hand experience. Rio was identified as a distant second, again pointing to the fact that the Rio Conference is probably the most famous and publicized event of the eight that featured in the survey. Although half the persons polled declined to put a number on the relevance of such declarations, here are some voices:

> They could be OK if they are followed. The delegations should involve other key players in the water sector and not only bureaucrats; some of them are just there by luck or political connections. (Anonymous)

Relevance is measured in the eye of the beholder and by what criteria observers use to judge them. The above quote is taken from a professional and reflects the concern of the people who have to carry through with implementation. Clearly in such a view the declarations become very unimportant if they have no perceivable effect on changing people's lives. As such, a few respondents made the connection between the declarations and subsequent local action. Two examples mentioned were the "Water for All and IDWSSD activities following the Mar del Plata Conservation of Environment policy" (Anonymous) in Zambia, and the Madagascar "national WASH campaign which came out of the World Summit" that stressed "the place of sanitation and hygiene promotion in water supply programmes" (Dorcas Pratt, Water Aid, Madagascar).

Now, consider the next statement that is made by a more political player:

> The whole continent of Africa is now talking and working hard to implement the Millennium Development Goals (MDG). (Anonymous)

This quote represents a very different view; one that focuses on articulating will and expressing ideas. Implementation comes second; and if not happening yet this is another issue. The question is therefore wrongly directed, and we should rather be asking: What realistic value do *we want* Ministerial Declarations to deliver?

12.3.6 Impact on the World of Water

Yes, megaconferences have had a marked effect on the water sector, at least if this is judged by the Southern African respondents. Table 8.3 shows that 72% thought the water sector would be different if these conferences had not been held, while 28% thought it would not have mattered. Broken down by number of conferences attended, there is an indication that those who have attended more conferences generally think they have also had an effect on the water sector. Hypothetically, attending more conferences makes it easier to make the link between subsequent impacts and a specific conference, of which attendees generally would have good knowledge. Conversely, non-attendance might make later changes hard to trace down to a specific event or conference (Table 12.3).

Table 12.3 The impact of megaconferences on the water sector according to number of conferences attended (percentage)

	Had impact	No impact
All	72	28
None	56	44
1–2	81	19
2+	100	0

These changes are connected to the specific strengths and weaknesses mentioned in a previous section, as well as to the original purposes of these conferences. Here are some comments:

The world of water has changed in that the political will for development and management of water resources in an integrated manner is taking off in most parts of the world. (Balisi Bernard Khupe, OKACOM, Botswana)

There is more awareness on the challenges at hand. Policy focus on the problems and needs has increased. (Patrick Okuni, Directorate of Water, Uganda)

Indeed the changes that have been perceived range from an increased political will, awareness and commitment globally, to providing specific targets, focus and funding for projects on "hot" topics.

We now have a target and focus for water related initiatives, but the primary change seems to be the flow of donor funds to specific projects. (Anonymous)

The conferences are also perceived to have brought new kinds of thinking into the water sector. Accurately or not several respondents associate the introduction of sustainable development and IWRM with specific conferences, changes that in their eyes would hardly have gained the kind of momentum they have without being aired at the global level.

12.3.7 Some Lessons Experience Has Taught Us

So what have we learned from the megaconferences that have been held? What messages can we bring to future events?

More than half the respondents thought the following statement best described their opinion:

C: The concept of such global conferences is good, but the present framework for organization needs to be changed radically. The events should be more focused and output-oriented. The main criteria for success should not be the number of people who attended the conference, but rather the quality of the results and their impacts.

And more than 25% thought the following was a fitting suggestion:

D: Instead of the global megaconferences it would be desirable to organize regional meetings, dealing with regional problems and issues, and which could be focused and impact-oriented.

Megaconferences in their present form certainly have not only their strengths, but also their weaknesses. Their sheer size and scale makes them useful tools for setting a global agenda, a feature that is, maybe, also their worst flaw. There is a strong sense in Southern Africa that they are "driven by developed countries' agendas" and "tend to be dominated by some groups" (Anonymous). Bluntly put,

They entrench the positions of the gate-keeping countries, institutions and elites. They are about recycling donor money back to donor countries. They are about a new form of chequebook diplomacy with specific objectives to be reached by the more powerful countries. (Anthony Turton, Gibb-SERA, South Africa)

That conferences are less about forging a common front against the global water crisis, and more about hidden agendas and financial interests, is a great blow to the future credibility and attractiveness of the events, as in the following quote regarding the World Summit on Sustainable Development in Johannesburg 2002:

> The exhibitions were totally dominated by business corporations and were quite frankly sterile. The Water Dome at the Jo'burg Summit was a deathly dull place! (Chris Dickens, Institute of Natural Resources, South Africa)

Many commentators feel that, with the exception of a few conferences, the major events create a buzz while they are happening, and then fade out over time. Issues discussed often end up in the same old debate, and several respondents feel that little new is produced when politics dominate the day.

The information that is generated is poorly distributed and often ends up on the bookshelves of those few who were there. Still, with the increasing use of the internet and other knowledge sharing tools, this is probably better than before – for those who have access to it. If well organized, the conferences provide powerful channels for knowledge sharing, networking and alignment of different world views. If conferences have taken a turn for the worse, there is a need for a debate on how to improve the situation. The current research is therefore timely and relevant.

12.4 Conclusion

From the answers gathered it can be deduced that some conferences are better than others. The quality, impact and usefulness, in the eyes of the respondents, seem to range from very poor to excellent or "revolutionary", with Rio as the most well-known and appreciated.

Megaconferences in their present form are partly suffering from a lack of credibility. They are sometimes seen as being increasingly favouring some groups or countries, with the polarization of developed and developing countries at the fore-front. Developing country representation, especially at the professional level, is also seen to be insufficient, partly due to the great costs associated with partaking in a major event. Such costs can often be ill-afforded by countries with strained resources.

While many see megaconferences as useful for aligning global efforts, raising awareness and sharing knowledge, the direct connection between the theoretical dimension and what is actually happening on the ground is less visible. The respondents to this survey do perceive that such practical results are lacking, but not totally absent. Quite a number of people said they knew of at least some outcomes of the mega events that eventually had a local and practical effect. Many questioned the link between related costs and apparent practical output however, as well as the local relevance of such events. There is no doubt that the functions respondents would like to see megaconferences perform are indeed driven by a specific need. We must ask ourselves if these global events really fulfil our needs in the best way possible, or if it would not be better to replace some of them with more regional forums – something that several people called for in their responses.

12.4.1 Voices from Southern Africa

What do people in Southern Africa in the water sector have to say about the impact of megaconferences on the water sector in general? In this section we let the respondents speak entirely for themselves with their own words. Here are two voices:

> Megaconferences perhaps have a value in bubbling the issues to the surface, and perhaps in building these issues into policies, targets and approaches. But I believe this has not lead to fundamental changes in the way developing nations function on the ground. If anything the conferences have only served to channel funding in different directions. (Anonymous)

> Information coming out of global megaconferences is often in hefty tomes. The usefulness of materials and documents needs to be given much more consideration. There is a place for heavy research documents—but there should also be more accessible user-friendly ways of communicating. This includes giving thought to the languages materials are available in as well as the layout. (Dorcas Pratt, Water Aid, Madagascar)

Acknowledgements

The authors would like to thank all of those who have put in valuable time and effort to express their views. Their contributions are greatly appreciated.

Appendix

A. Questionnaire

Impacts of Global Megaconferences on Water

Conducted by the African Water Issues Research Unit (University of Pretoria) on behalf of the Third World Centre for Water Management, Mexico.

This questionnaire can be returned to us anonymously by e-mail (MDrackner@csir.co.za) or by fax: +27(0)866-725962 as soon as possible. We are soliciting views from those who have attended one or more of these megaconferences, as well as from those who have not attended any of these conferences. If acceptable to you, we would prefer to receive your comments formally for possible future interactions. Should you agree to this request, we wish to assure you that we shall NOT attribute any comments to you, without your explicit authorization. Thank you for your time.

Please be frank in your statements: "politically correct" views are unlikely to be of much use in this assessment.

1. Your Participation – Mark with X. Did you participate in:

Conference

- UN Water Conference, Mar del Plata, 1977
- International Conference on Water and the Environment, Dublin, 1992
- UN Conference on Environment and Development, Rio de Janeiro, 1992
- First World Water Forum, Marrakech, 1997
- Second World Water Forum, the Hague, 2000
- International Conference on Freshwater, Bonn, 2001
- UN Conference on Sustainable Development, Johannesburg 2002
- Third World Water Forum, Japan, 2003

2. Your overall views on each megaconference – Based on your current knowledge of these megaconferences (whether you participated or not), please state your overall views on each of the event(s) in a scale of 0 (very poor) – 5 (absolutely excellent), based on your own perception of their outputs and impacts. Use 3 for average. If you have no specific knowledge on a conference, please say N/A.

Conference

- UN Water Conference, Mar del Plata, 1977
- International Conference on Water and the Environment, Dublin, 1992
- UN Conference on Environment and Development, Rio de Janeiro, 1992
- First World Water Forum, Marrakech, 1997
- Second World Water Forum, the Hague, 2000
- International Conference on Freshwater, Bonn, 2001
- UN Conference on Sustainable Development, Johannesburg 2002
- Third World Water Forum, Japan, 2003

3. Impacts of megaconferences – Irrespective of whether you participated or not in these megaconferences, please select which one of the following comments most closely reflects your overall views on all the megaconferences as a whole.

A. The conferences were excellent. They have radically increased my knowledge-base, and have improved my working practices significantly.

B. The conferences have significantly changed the policies, programmes and projects of my institution. These changes would not have happened if these conferences had not taken place.

C. The conferences had at best a marginal impact on me and/or my institution.

D. The conferences had no perceptible impact on me and/or my institution.

E. It was pleasant to attend the conference(s), meet old friends, and make new ones, but the conferences really had no lasting or visible impacts on me and/or my institution.

F. These were mostly forgettable events. For all practical purposes, it would not have mattered much whether these events had ever been held or not. They simply did not leave any footprints on water management.

4. Strengths of megaconferences – What in your view are the most important strengths of these megaconferences? (List maximum three strengths.)

5. Weaknesses of megaconferences – What in your view are the most important weaknesses of these megaconferences? (List maximum three weaknesses.)

6. Cost-effectiveness of megaconferences. The mega conferences are often expensive to organize, and the costs seem to have increased significantly in recent years. For example, the costs of organizing the UN Water conference in Mar del Plata or the First World Water Forum in Marrakech were modest. The cost of organizing the Second World Water Forum was much higher. The cost of the Third World Water Forum was significantly higher than the Hague Forum. The cost of the Secretariats alone for the Forums are quite high: normally well over $10 million.

Based on your perception of their outputs and impacts, what is your view of the cost-effectiveness of these events (0–5)?

5 Extremely high
4 High
3 Moderate
2 Low
1 Extremely low
0 None

7a. Documentation and information dissemination – Do you have adequate documentation (reports, papers, proceedings, etc.) from any of these conferences?

Yes
No

7b. If yes, which ones? How useful has this documentation been (0–5)?

Conference	Type of documentation	Usefulness (0–5)
• UN Water Conference, Mar del Plata, 1977		
• International Conference on Water and the Environment, Dublin, 1992		
• UN Conference on Environment and Development, Rio de Janeiro, 1992		
• First World Water Forum, Marrakech, 1997		
• Second World Water Forum, the Hague, 2000		
• International Conference on Freshwater, Bonn, 2001		
• UN Conference on Sustainable Development, Johannesburg 2002		
• Third World Water Forum, Japan, 2003		

7c. Overall, how do you rate the quality of documents you have seen of these megaconferences (0–5)?

5 Extremely high
4 High
3 Moderate
2 Low
1 Extremely low
0 None

8a. Practical results from the megaconferences – In your view, did any of these megaconferences yield positive, implementable and lasting results?

Yes
No

8b. In your view, did any new initiatives (including water sector reform) originate from these events, which otherwise would not have occurred?

Yes
No

8c. Are you aware of any policy change in your country which would not have occurred without one or more of these megaconferences?

Yes
No

8d. If yes on any of questions 8a–c, please give examples from the specific megaconferences at regional, national and/or global events (practical results, new initiatives, policy changes).

Conference	Examples
• UN Water Conference, Mar del Plata, 1977	
• International Conference on Water and the Environment, Dublin, 1992	
• UN Conference on Environment and Development, Rio de Janeiro, 1992	
• First World Water Forum, Marrakech, 1997	
• Second World Water Forum, the Hague, 2000	
• International Conference on Freshwater, Bonn, 2001	
• UN Conference on Sustainable - Development, Johannesburg 2002	
• Third World Water Forum, Japan, 2003	

9. Changes in investment due to megaconferences – Has the investments availability for the water sector in your country increased or decreased due to these conferences in a way that would not have occurred unless these conferences had taken place? If it has changed, please give the direction and a rough estimate of these changes

	It has increased	It has decreased	No change
Direction			
Estimate of change			

10. Key Lessons – What in your view are the key lessons (positive and negative) that we can learn from these megaconferences?

	Positive	Negative
Key Lesson 1		
Key Lesson 2		
Key Lesson 3		

11a. Changes in the world of water due to megaconferences – In your view, would the world of water have been any different if these conferences had not taken place?

Yes
No

11b. If yes, in what ways has it changed?

Changes

12. Your overall view on the megaconferences – What is your overall view of these megaconferences? Choose the statement that is closest to your view.

A. The global megaconferences are useful and cost-effective. We should continue with them with few changes

B. The conferences have now become one big "water fair", with a lot of activities but without much thought as to their relevance, appropriateness, outputs or impacts. There is no coordination between events, no clear focus, and their cost-effectiveness leaves much to be desired.

C. The concept of such global conferences is good, but the present framework for organization needs to be changed radically. The events should be more focused and output-oriented. The main criteria for success should not be the number of people who attended the conference, but rather the quality of the results and their impacts.

D. Instead of the global megaconferences it would be desirable to organize regional meetings, dealing with regional problems and issues, and which could be focused and impact-oriented.

13. Ministerial Declarations – At many of these conferences, there were Ministerial Declarations which had relevance to the water sector. Please give your opinions on the following questions.

13a. Are you and your colleagues aware of these Ministerial Declarations?

Yes
No

13b. If yes, please identify the conference whose declaration you consider had the most impact and was best

Conference
• UN Water Conference, Mar del Plata, 1977
• International Conference on Water and the Environment, Dublin, 1992
• UN Conference on Environment and Development, Rio de Janeiro, 1992
• First World Water Forum, Marrakech, 1997
• Second World Water Forum, the Hague, 2000
• International Conference on Freshwater, Bonn, 2001
• UN Conference on Sustainable Development, Johannesburg 2002
• Third World Water Forum, Japan, 2003

13c. What, in your view, is the relevance, appropriateness and usefulness of such declarations (0–5)?

5 Extremely high
4 High
3 Moderate
2 Low
1 Extremely low
0 None

13d. Has the water policy and/or priorities of water programmes of your country been affected by these Ministerial Declarations?

Yes
No

13e. If yes, please briefly provide examples:

14. Additional – Please give your views on any other aspect(s) and issue(s) of global megaconferences not mentioned above. This could be as long as you like:

15. Details – the completion of this section is optional – we shall not use any one's name in the evaluation report without explicit authorization from the individual concerned:

Name:	Organization:	
E-mail:	Tel:	
I agree to the formal use of my comments:	Yes	No

B. Spatial Distribution of Responses

Country	Number of persons directly polled*	Number of responses	Response rate (%)
South Africa	53	15	28
Zimbabwe	25	2	8
Namibia	18	3	17
Mozambique	8	1	12.5
Madagascar	4	1	25
Angola	0	0	–
Uganda	9	1	11
Kenya	10	0	0
Tanzania	13	0	0
Botswana	13	4	31
Swaziland	4	1	25
Lesotho	6	1	17
Zambia	20	1	5
Malawi	8	0	0
Seychelles	1	0	0
Mauritius	1	0	0
DRC	1	0	0

* This number signifies individuals on the original contact list. Included are also people whose e-mail addresses failed during the original send-out or during one of the subsequent reminders.

NA-D

Index

N

O

P

T

U